biology NOW!

11-14

second edition

D0911791

98413 55037 — Padmavathi Nagar
98413 55056

biology NOW!

11-14

second edition

Peter D Riley

HODDER
EDUCATION
AN HACHETTE UK COMPANY

Titles in this series

Biology Now! 11–14 Second Edition Pupil's Book   ISBN 978 0 71958 060 4
Biology Now! 11–14 Second Edition
   Teacher's Resource Book   ISBN 978 0 71958 059 8
Chemistry Now! 11–14 Second Edition Pupil's Book ISBN 978 0 71958 062 8
Chemistry Now! 11–14 Second Edition
   Teacher's Resource Book   ISBN 978 0 71958 061 1
Physics Now! 11–14 Second Edition Pupil's Book ISBN 978 0 71958 064 2
Physics Now! 11–14 Second Edition
   Teacher's Resource Book   ISBN 978 0 71958 063 5

Although every effort has been made to ensure that website addresses are correct at the time of going to press, Hodder Education cannot be held responsible for the content of any website mentioned in this book.

Hachette's policy is to use papers that are natural, renewable and recyclable products and made from wood grown in sustainable forests. The logging and manufacturing processes are expected to conform to the environmental regulations of the country of origin.

Orders: please contact Bookpoint Ltd, 130 Milton Park, Abingdon, Oxon OX14 4SB. Tel (44) 01235 827720. Fax (44) 01235 400454. Lines are open from 9.00–5.00, Monday to Saturday, with a 24-hour message answering service. Visit our website *www.hoddereducation.co.uk*

© Peter Riley 2004

First published in 2004
by Hodder Education, an Hachette UK Company
338 Euston Road
London NW1 3BH

Reprinted 2000, 2001 (with revisions)
Second edition 2004, reprinted 2005, 2008 (twice), 2009

All rights reserved. No part of this publication may be reproduced in any material form (including photocopying or storing in any medium by electronic means and whether or not transiently or incidentally to some other use of this publication) without the written permission of the Publisher, except in accordance with the provisions of the Copyright, Designs and Patents Act 1988 or under the terms of a licence issued by the Copyright Licensing Agency.

Artwork by Mike Humphries and Linden Artists
Cover design by John Townson/Creation

Typeset in 12/14 Garamond by Pantek Arts Ltd

Printed and bound in Italy

A CIP catalogue record for this book is available from the British Library

ISBN: 978 0 71958 060 4

Teacher's Resource Book ISBN 978 0 71958 059 8

Contents

Preface

To the pupil

Biology is the scientific study of living things. It includes investigations on tiny structures, such as cells, and on huge structures, such as a rainforest, an ocean or even the whole Earth! Some biologists are even looking for signs of life in other parts of the Solar System or on planets around other stars.

Our knowledge of biology has developed from the observations, investigations and ideas of many people over a long period of time. Today this knowledge is increasing more rapidly as there are more biologists – people who study living things – than ever before.

In the past, few people other than scientists were informed about the latest discoveries. Today, through newspapers, television and the internet, everyone can learn about the latest discoveries on a wide range of biological topics, from curing illnesses and developing new foods to ways of reducing environmental damage and conserving rare species.

Biology Now! 11–14 covers the requirements of your examinations in a way that I hope will help you understand how observations, investigations and ideas have led to the scientific facts we use today. The questions are set to help you extract information from what you read and see, and to help you think more deeply about each chapter in this book. Some questions are set so you can discuss your ideas with others and develop a point of view on different scientific issues. This should help you in the future when new scientific issues, which are as yet unknown, affect your life.

The scientific activities of thinking up ideas to test and carrying out investigations are enjoyed so much by many people that they take up a career in science. Perhaps *Biology Now! 11–14* may help you to take up a career in science too.

To the teacher

The first edition of *Biology Now! 11–14* was written to cover the requirements of the curriculum for the Common Entrance Examination at 13+, the National Curriculum for Science at Key Stage 3, and equivalent junior courses. It was published before the QCA Scheme for Science at Key Stage 3. This new edition has been prepared with the QCA scheme in mind, but it also holds to the aims of the previous edition. The aims are:

- to help pupils become more scientifically literate by encouraging them to examine the information in the text and illustrations in order to answer questions about it in a variety of ways. For example, *For discussion* questions may be used in work on science and citizenship
- to present science as a human activity by considering the development of scientific ideas from the earliest times to the present day
- to examine applications of scientific knowledge, and the issues that arise from them.

The chapters in this second edition are arranged to follow the QCA units for biology in Years 7, 8 and 9. The book is supported by a *Teacher's Resource Book* that provides answers to all the questions in the pupils' books – those that occcur in the body of the chapter, and those that occur as end-of-chapter tests and which may be used for assessment.

The resource book also provides extra material for use in assessment in the form of end-of-chapter tests, a bank of questions in the style of the 13+ examination questions and actual questions from the past papers of the Key Stage 3 examination. The *Teacher's Resource Book* begins with a bridging unit. This relates to the first chapter of the Pupil's Book and may help the pupils settle in and let them carry forward some of their studies at primary school into a secondary school laboratory setting. This is followed by chapter by chapter support which includes information to help you track the National Strategy for Science objectives and suggestions for several lesson starters with page suggestions where they may be used. There is also a range of practical activities for integration with the work to provide opportunities for pupils to develop their skills in scientific investigation.

Although *Biology Now! 11–14* second edition was written to provide the biology content of a balanced science course in which biology, chemistry and physics are taught separately, it may also be used as a supplementary text in more integrated courses to demonstrate aspects of science as a human activity, and to extend skills in comprehension.

Acknowledgements

To Anita

I would like to thank Katie Mackenzie Stuart and Amy Austin for their help in the preparation of this second edition of *Biology Now! 11–14*.

The cover photograph is of lepidopteron wing scales.

Thanks to FRANK for use of the drugs posters in this book. FRANK is available 24 hours every day for confidential advice and information about drugs, 0800 77 66 000 or visit talktofrank.com.

Cover Eye of Science/Science Photo Library; **p.1** *tl & r* Heather Angel, *bl* Gerard Lacz/NHPA; **p.8** J.C. Revy/Science Photo Library; **p.12** Carolina Biological Supply Company/Oxford Scientific Films; **p.15** Manfred Kage/Science Photo Library; **p.18** *t & b* Eye of Science/Science Photo Library; **p.21** Stephen Dalton/NHPA; **p.28** Jurgen Freund/Naturepl; **p.31** Mike James/Science Photo Library; **p.32** S.I.U. School of Medicine/Science Photo Library; **p.33** Andrew Lambert; **p.34** © Bubbles/John Garrett; **p.39** CNRI/Science Photo Library; **p.40** Motta & Familiari/Anatomy Dept./University "La Sapienza", Rome/Science Photo Library; **p.41** *l* Tim Beddow/Science Photo Library, *r* Hank Morgan/Science Photo Library; **p.42** *l* Sally Greenhill © Sally & Richard Greenhill, *r* Andrew Lambert; **p.47** Stevie Grand/Science Photo Library; **p.48** Sally Greenhill © Sally and Richard Greenhill; **p.50** David Woodfall/NHPA; **p.52** David Woodfall/NHPA; **p.54** John Hawkins/Frank Lane Picture Agency; **p.55** Stephen Dalton/NHPA; **p.57** *l* © Neil McIntyre, *r* Allan G. Potts/Bruce Coleman Ltd; **p.59** *t* G.I. Bernard/NHPA, *b* Heather Angel; **p.65** G.I. Bernard/NHPA; **p.66** *l* Heather Angel, *c* Andrew Henley/Natural Visions, *r* Jany Sauvanet/NHPA; **p.67** Sally Greenhill © Sally and Richard Greenhill; **p.69** © Breck P. Kent/Earth Scenes/Oxford Scientific Films; **p.73** *l* Michele Hall/Oxford Scientific Films, *tr* A.N.T./NHPA, *br* Jeff Foot Productions/Bruce Coleman Ltd; **p.80** Andrew Lambert; **p.82** John Townson/Creation; **p.83** Biophoto Associates/Science Photo Library; **p.85** Heather Angel; **p.91** *t* Royal College of Physicians Photo Library, *b* Jason Venus/Natural Visions; **p.107** Biophoto Associates/Science Photo Library; **p.111** Wellcome Institute Library, London; **p.116** *t* Heather Angel, *b* John Townson/Creation; **p.123** WaterAid/Caroline Penn; **p.127** © The Trustees of the Wellcome Trust/National Medical Slide Bank; **p.130** *t* Tim Beddow/Science Photo Library, *b* Medical Illustration Services/Glasgow Royal Infirmary University NHS Trust; **p.132** *t* John Durham/Science Photo Library, *b* Mary Evans Picture Library; **p.136** *t* © Derek Whitford, *b* David Woodfall/NHPA; **p.137** Harwood/Ecoscene; **p.138** Sally Morgan/Ecoscene; **p.139** © Jacomina Wakeford/ICCE; **p.147** © Richard Davies/Oxford Scientific Films; **p.151** © Wildlife Matters; **p.152** *t* Alexandra Jones/Ecoscene, *b* © Garden & Wildlife Matters; **p.153** John Mason/Ardea London Ltd; **p.159** A.B. Dowsett/Science Photo Library; **p.165** Associated Press; **p.166** Simon Fraser/Science Photo Library; **p.167** John Townson/Creation; **p.168** *l* Holt Studios/Bob Gibbons, *r* Holt Studios/Nigel Cattlin; **p.175** Deep Light Productions/Science Photo Library; **p.181** Damien Lovegrove/Science Photo Library; **p.187** FRANK (leaflets photographed by John Townson/Creation); **p.188** FRANK (photographed by John Townson/Creation); **p.192** Heather Angel; **p.198** Harry Smith Horticultural Photographic Collection; **p.201** *all* Holt Studios/Nigel Cattlin; **p.205** Stefan Meyers/Ardea London Ltd; **p.207** *all* John Townson/Creation; **p.212** Holt Studios/Nigel Cattlin; **p.215** © Underwood & Underwood/Corbis; **p.218** *both* John Townson/Creation; **p.220** Holt Studios/Primrose Peacock.

t = top, *b* = bottom, *l* = left, *r* = right, *c* = centre

Every effort has been made to contact copyright holders but if any have been inadvertently overlooked the Publishers will be pleased to make the necessary arrangements at the earliest opportunity.

1 The structure of living things

Figure 1.1 There is a great variety in the structure of living things, as these photographs show.

Organs and organ systems

Biology is the study of living things. The parts of a body that perform tasks to keep a living thing alive are called organs and most organs work together in groups called organ systems.

Organ systems of a human

There are ten organ systems in the human body. They are listed below but will be covered in more detail throughout the book. The tasks they carry out are sometimes called life processes.

1   The **sensory system** is made up of sense organs such as the eye and the ear. The function of this system is to provide information about the surroundings.

1

1 Which organ system:
  a) transports materials around the body
  b) absorbs food into the blood
  c) detects changes in the environment
  d) produces hormones
  e) co-ordinates activities
  f) takes in oxygen from the air
  g) supports the body
  h) produces offspring
  i) removes waste from the blood
  j) moves bones?

2 Which organ system or systems are involved in:
  a) movement
  b) nutrition
  c) circulation?

3 What other sense organs are in the sensory system?

4 What is the name of the hormone that makes your heart beat faster and directs more blood to your muscles?

5 What movements take place in the body that you do not have to think about?

6 How does your pattern of breathing change when you exercise and then rest?

7 Write down the names of any bones that you know (without looking them up) and say where they are found in the body.

2 The **nervous system** comprises the brain, spinal cord and nerves. This system controls the actions of the body and co-ordinates many of its activities without you having to think about them. For example, you breathe in and out automatically.

3 The **respiratory system** is located in the chest. It is formed by the windpipe, the lungs, the ribs and rib muscles (called the intercostal muscles) and the diaphragm. The system works to draw in air and then expel it. While in the lungs oxygen passes from the air into the blood and carbon dioxide passes from the blood into the air.

4 The **digestive system** is a long tube through the body in which food is broken down and absorbed into the blood. It is made up of many organs including the liver, which also performs many other tasks in the body.

5 The **circulatory system** transports materials around the body in a liquid called the blood. Blood is moved along tubes called blood vessels by the pumping action of the heart.

6 The **excretory system** cleans waste from the blood by a filtration process in the kidneys. The liquid containing the waste is called urine. It is stored in the bladder before it is released.

7 The **skeletal system** is made up of 206 bones. They provide support for the body and have joints between them that help the body to move. Some bones form a protective structure, for example, the bones in the skull form a protective case around the brain.

8 The **muscle system** provides the mechanism for movement. A muscle is capable of making itself shorter to exert a pulling force on a bone.

9 The **endocrine system** is made up of glands which release chemicals called hormones into the blood. The adrenal gland is an example of an endocrine gland. It is found just above the kidney and releases (or secretes) a hormone called adrenaline. You feel the effect of adrenaline if you are asked to read aloud or act in front of a large audience, or take part in athletics. It makes your heart beat faster and directs more blood to your muscles.

10 The **reproductive system** of the male produces sperm cells and the reproductive system of the female produces eggs and provides a place for a baby to grow.

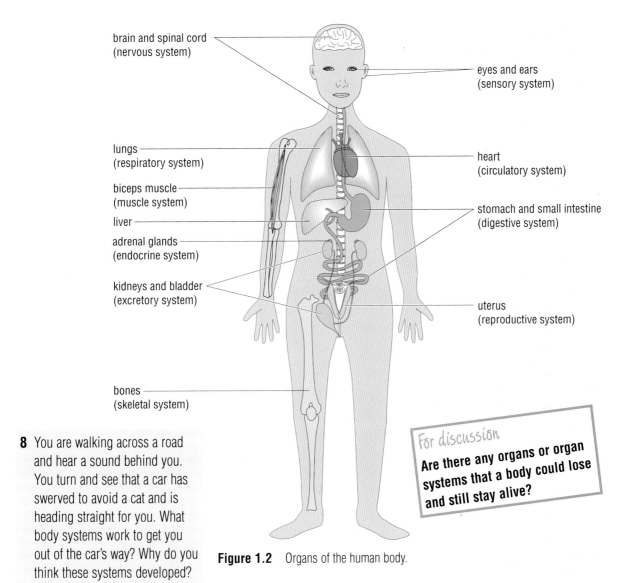

brain and spinal cord
(nervous system)

eyes and ears
(sensory system)

lungs
(respiratory system)

heart
(circulatory system)

biceps muscle
(muscle system)

liver

stomach and small intestine
(digestive system)

adrenal glands
(endocrine system)

kidneys and bladder
(excretory system)

uterus
(reproductive system)

bones
(skeletal system)

8 You are walking across a road and hear a sound behind you. You turn and see that a car has swerved to avoid a cat and is heading straight for you. What body systems work to get you out of the car's way? Why do you think these systems developed?

Figure 1.2 Organs of the human body.

For discussion

Are there any organs or organ systems that a body could lose and still stay alive?

Organs of a flowering plant

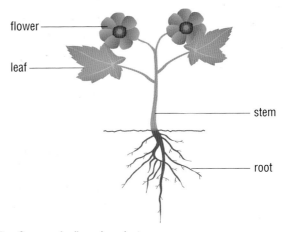

flower

leaf

stem

root

Figure 1.3 Organs of a flowering plant.

9 Draw a table featuring the organs of a flowering plant and the tasks they perform.

10 How is the leaf dependent on the root and the stem?

11 Which life processes or tasks do you think are found in both plants and humans? Explain your answer.

There are four main organs in the body of a flowering plant. They are the root, stem, leaf and flower. Each organ may be used for more than one task or life process.

1 The **root** anchors the plant and takes up water and minerals from the soil. The roots of some plants, such as the carrot, store food.

2 The **stem** transports water and food and supports the leaves and the flowers. Some plants, such as trees, store food in their stems.

3 The **leaf** produces food. In some plants, such as the onion, food is stored in the bases of the leaves. The swollen leaf bases make a bulb.

4 The **flower** contains the reproductive organs of the plant.

All the organs work together to keep the plant alive so that it can grow and produce offspring.

From organs to cells

Marie F. X. Bichat (1771–1802) was a French doctor who did many post mortems. In the last year of his life he carried out 600. He cut up the bodies of dead people to find out how they had died. From this he discovered that organs were made of layers of materials. He called these layers 'tissues' and identified 21 different kinds. For a while scientists thought that tissues were made of simple non-living materials.

In 1665, long before Bichat was born, an English scientist named Robert Hooke (1635–1703) used a microscope to investigate the structure of a very thin sheet of cork. He discovered that it had tiny compartments in it. He thought of them as rooms and called them 'cells', after the small rooms in monasteries where monks worked and meditated.

Bichat did not examine the tissues he had found under a microscope because most of those made at that time did not produce very clear images. When better microscopes were made, scientists investigated pieces of plants and found that, like cork, they also had a cell structure. The cells in Hooke's piece of cork had been empty but other plant cells were found to contain structures.

A Scottish scientist called Robert Brown (1773–1858) studied plant cells and noticed that each one had a dark spot inside it. In 1831 he named the spot the 'nucleus' which means 'little nut'.

Matthias Schleiden (1804–1881) was a German scientist who studied the parts of many plants. In 1838 he put forward a theory that all plants were made of cells. A year later Theodor Schwann, another German scientist, stated that animals were also made of cells.

The ideas of Schleiden and Schwann became known as the Cell Theory. It led other scientists to make more discoveries about cells and to show that tissues are made up of groups of similar cells.

1 Where did Bichat get his ideas that organs were made from tissues?

2 Who first described 'cells' and where did the idea for the word come from?

3 Who named the nucleus and what does it mean?

4 What instrument was essential for the study of cells?

5 How could the Cell Theory have been developed sooner?

6 Arrange these parts of a body in order of size starting with the largest: cell, organ, tissue, organ system.

The microscope

A microscope is used for looking at specimens very closely. Most laboratory microscopes give a magnification up to about 200 times but some can give a magnification of over 1000 times. The microscope must also provide a clear view and this is achieved by controlling the amount of light shining onto the specimen.

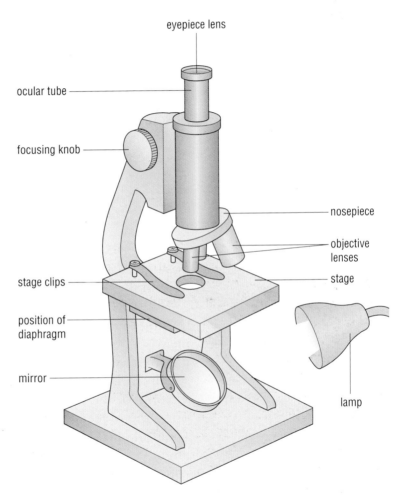

Figure 1.4 The main parts of a microscope.

Light is collected by a mirror at the base of the microscope. The mirror is held in special joints that allow it to move in any direction. The light comes from a lamp or from a sunless sky. It must never be collected directly from the Sun as this can cause severe eye damage and blindness. Some microscopes have a built-in lamp instead of a mirror. The light either shines directly through a hole in the stage onto the specimen or it passes through a hole in a diaphragm. The

12 What is a microscope used for?

13 What advice would you give someone about how to collect light to shine into a microscope?

14 What magnification would you get by using an eyepiece of ×5 magnification with an objective lens of ×10 magnification?

15 If you had a microscope with ×5 and ×10 eyepieces and objective lens of ×10, 15 and 20, what powers of magnification could your microscope provide?

16 How would you advise someone to use the three objective lenses on the nosepiece?

17 Why should you not look down the microscope all the time as you try to focus the specimen?

18 Look at the picture of the microscope on page 5 and describe the path taken by light from a lamp near the microscope to the eye.

diaphragm allows the amount of light reaching the specimen to be controlled by increasing or decreasing the size of the hole. The light shining through a specimen is called transmitted light.

Above the specimen is the ocular tube. This has an eyepiece lens at the top and one or more objective lenses at the bottom. The magnification of the two lenses is written on them. An eyepiece lens may give a magnification of ×5 or ×10. An objective lens may give a magnification of ×10, 15 or 20. The magnification provided by both the eyepiece lens and the objective lens is found by multiplying their magnifying powers together. Most microscopes have three objective lenses on a nosepiece at the bottom of the ocular tube. The nosepiece can be rotated to bring each objective lens under the ocular tube in turn. An investigation with the microscope always starts by using the lowest power objective lens then working up to the highest power objective lens if it is required.

A specimen for viewing under the microscope must be put on a glass slide. The slide is put on the stage and held in place by the stage clips. The slide should be positioned so that the specimen is in the centre of the hole in the stage.

The view of the specimen is brought into focus by turning the focusing knob on the side of the microscope. This may raise or lower the ocular tube or it may raise or lower the stage on which the slide of the specimen is held. In either case you should watch from the side of the microscope as you turn the knob to bring the objective lens and specimen close together. If you looked down the ocular tube as you did this you might crash the objective lens into the specimen which could damage both the lens and the specimen. When the objective lens and the specimen are close together, but not touching, look down the eyepiece and turn the focusing knob so that the objective lens and specimen move apart. If you do this slowly, the blurred image will become clear.

Finding the size of microscopic specimens

The disc of light you see when you look down a microscope is called the field of view. You can estimate the size of the specimens you see under the microscope if you know the size of the field of view. A simple way to find the size of the field of view is to put a piece of

19 A field of view was found to be 2000 µm in diameter. A soil particle reached one quarter of the way across it. How long would you estimate the length of the particle to be?

20 The fields of view of three lenses were measured. A was 100 µm, B was 3000 µm and C was 500 µm. Which was the most powerful lens and which was the least powerful lens?

21 Why does the field of view decrease as the power of the objective lens increases?

graph paper on a slide and examine it using the low power objective lens. The squares on the graph should be 1 mm across. Microscopic measurements are not made in millimetres, they are measured in micrometres. 1 mm = 1000 micrometres (written as 1000 µm).

If the field of view is two squares across it has a diameter of 2000 µm. If you remove the slide with the graph paper and replace it with a slide with some soil particles, you could estimate the size of a soil particle by judging how far it crosses the field of view.

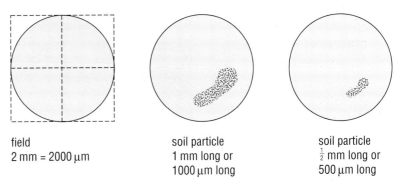

field
2 mm = 2000 µm

soil particle
1 mm long or
1000 µm long

soil particle
$\frac{1}{2}$ mm long or
500 µm long

Figure 1.5 A soil particle under the microscope.

If the soil particle comes halfway across the field of view it is 1000 µm long. There is a relationship between the power of an objective lens and its field of view. As the power of the objective lens increases, the size of its field of view decreases.

Cells

There are ten times more cells in your body than there are people on the Earth. If you stay in the water a long time at a swimming pool you may notice that part of your skin sometimes flakes off when you dry yourself. These flakes are made of dead skin cells. You are losing skin cells all the time but in a much smaller way. As your clothes rub against your skin they pull off tiny flakes which pass into the air and settle in the dust. A small part of the dirt that cleaners sweep up at the end of a school day comes from the skin that the pupils have left behind.

Figure 1.6 shows a section of human skin that has been stained and photographed down a microscope using a high power objective lens. When unstained, the different parts of the cells are colourless and are difficult to distinguish. In the 1870s it was discovered that dyes could be made from coal tar which would stain different parts of the cell. Cell biologists found they could stain the nucleus and other parts of the cell different colours to see them more easily.

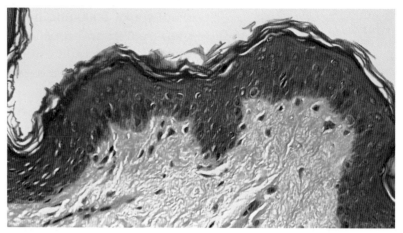

Figure 1.6   Section of human skin. Cells can be seen flaking off the surface.

22 Why are most specimens of cells stained before they are examined under the microscope?

23 You look down a microscope at a slide labelled 'Cells'. You can see a coloured substance with dots in it and lines that divide the substance into rectangular shapes. Inside the rectangular shapes, what are:
 a) the dots
 b) the lines
 c) the coloured substance?

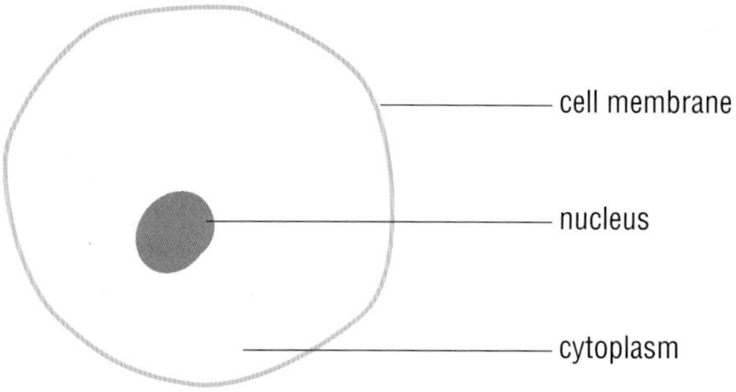

cell membrane

nucleus

cytoplasm

Figure 1.7   A typical animal cell.

Basic parts of a cell

Nucleus

This is the control centre of the cell. It contains the genetic material, called DNA (its full name is deoxyribonucleic acid). The DNA molecule is a long chain of smaller molecules. They occur in different combinations along the DNA molecule. The combinations of molecules provide instructions for the cell to make chemicals to keep it alive or to build its cell parts. When a cell divides the DNA divides too, so that the nucleus of each new cell receives all the instructions to keep the new cell alive and enable it to grow.

24 How does the cell membrane protect the cell?

25 If there are about 6000 million people on the Earth, how many cells have you got in your body?

Cytoplasm

This is a watery jelly which fills most of the cell in animal cells. It can move around inside the cell. The cytoplasm may contain stored food in the form of grains. Most of the chemical reactions that keep the cell alive take place in the cytoplasm.

Cell membrane

This covers the outside of the cell and has tiny holes in it called pores that control the movement of chemicals in or out of the cell. Dissolved substances such as food, oxygen and carbon dioxide can pass through the cell membrane. Some harmful chemicals are stopped from entering the cell by the membrane.

Parts found only in plant cells

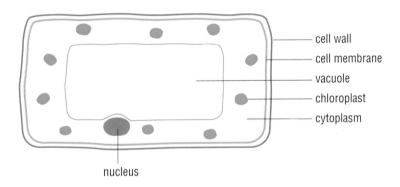

Figure 1.8 A typical plant cell.

Cell wall

This is found outside the membrane of a plant cell. It is made of cellulose which is a tough material that gives support to the cell.

Chloroplasts

These are found in the cytoplasm of many plant cells. They contain a green pigment called chlorophyll which traps a small amount of the energy in sunlight. This energy is used by the plant to make food in a process called photosynthesis (see Chapter 11). Chloroplasts are found in many leaf cells and in the stem cells of some plants.

26 Name two things that give support to a plant cell.

27 Would you expect to find chloroplasts in a root cell? Explain your answer.

28 Why do plants wilt if they are not watered regularly?

Large vacuole

This large space in the cytoplasm of a plant cell is filled with a liquid called cell sap which contains dissolved sugars and salts. When the vacuole is full of cell sap the liquid pushes outwards on the cell wall and gives it support. If the plant is short of water, the support is lost and the plant wilts.

Some animal cells and Protoctista (see page 120) have vacuoles but they are much smaller than those found in plant cells.

Adaptation in cells

The word adaptation means the change of an existing design for a particular task (see also page 51). The basic designs of plant and animal cells were shown in the last section, but many cells are adapted which allows them to perform a more specific task. Here are some common examples of the different types of plant and animal cells.

Root hair cells

These grow a short distance behind the root tip. The cells have long thin extensions that allow them to grow easily between the soil particles. The shape of these extensions gives the root hair cells a large surface area through which water can be taken up from the soil.

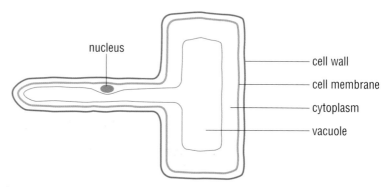

Figure 1.9 A root hair cell.

Palisade cells

These plant cells have a shape that allows them to pack closely together in the upper part of a leaf, near the light. They have large numbers of chloroplasts in them to trap as much light energy as possible.

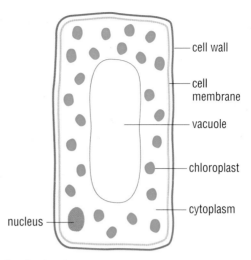

Figure 1.10 A palisade cell.

Ciliated epithelial cells

Cells that line the surface of structures are called epithelial cells. Cilia are microscopic hair-like extensions of the cytoplasm. If cells have one surface covered in cilia they are described as ciliated. Ciliated epithelial cells line the throat. Air entering the throat contains dust that becomes trapped in the mucus of the throat lining. The cilia wave to and fro and carry the dust trapped in the mucus away from the lungs.

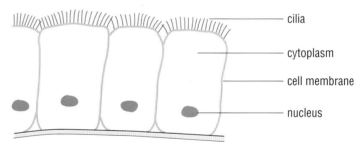

Figure 1.11 Ciliated epithelial cells.

Sperm cells

These transport the male genetic material in their nucleus. They have a streamlined shape which allows them to move easily through the liquid as they travel towards the female egg (see Figure 1.12 overleaf). They have a tail that waves from side to side to push the cell forwards.

29 What changes have taken place in the basic plant cell to produce a root hair cell?

30 How is a palisade cell different from a root hair cell? Explain these differences.

31 Why would it be a problem if root hair cell extensions were short and stubby?

32 How are sperm and egg cells
  a) similar and
  b) different?

33 Smoking damages the cilia lining the breathing tubes. What effect might this have on breathing?

34 Why are there different kinds of cells?

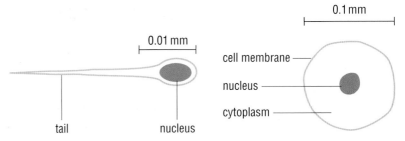

Figure 1.12 Sperm and egg cell (showing size scale).

Egg cells

These contain the female genetic material in their nucleus. They are much larger than sperm cells because they contain a food store, and they do not move on their own. When a sperm reaches an egg the genetic material from the sperm combines with that in the egg in a process called fertilisation. This produces a cell which divides many times to produce an embryo. The food and energy for early growth of the embryo is provided by the egg.

From cells to tissues

Cells are not mixed up inside the body. They are arranged in an orderly way in groups. A group of cells of the same type is called a tissue, and it performs an important task in the life of an organism. Figure 1.6 on page 8 shows a tissue of cells. They are called epithelial cells and their task is to make a waterproof and germ-proof covering over the body. Figure 1.11 on page 11 shows some cells from a tissue in the throat. The function of this tissue is to transport harmful particles away from the lungs.

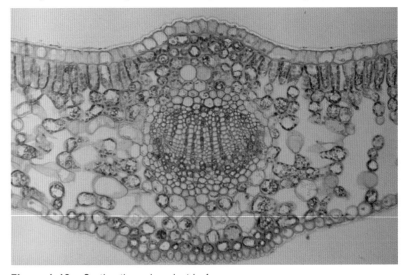

Figure 1.13 Section through a plant leaf.

35 What is a tissue?

36 a) Name the tissues labelled A, B and C in the section of this leaf.

b) What is X?

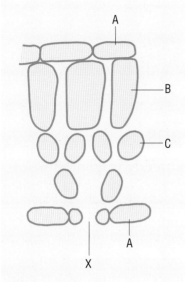

Figure 1.14

37 a) Compare the tissue of the epidermis and the palisade tissue of the leaf.

b) How are the tasks of the palisade tissue and the spongy mesophyll
i) similar and
ii) different?

Tissues in plants

If you examine a very thin slice of a leaf (called a section) under the microscope you can see how it is composed of different tissues of cells (see Figure 1.13).

Covering tissue

The surface of the leaf is covered with a tissue of flat, transparent cells which make up the epidermis. This surface prevents water escaping from the leaf or from entering it. It also allows light rays to pass through the tissues of cells inside the leaf.

Energy-trapping tissue

Under the upper epidermis is the palisade tissue, made from tightly packed palisade cells (see Figure 1.10, page 11). These cells contain many chloroplasts and trap some of the energy in sunlight to make food (see Chapter 11).

Spongy tissue

Below the palisade tissue is a tissue called the spongy mesophyll. The cells in this tissue can also collect energy in sunlight and make food but their main purpose is to provide a large surface area inside the leaf from which water can evaporate. The evaporation of water inside the leaf has two purposes. Firstly, when water evaporates it takes heat energy from its surroundings. This helps to cool the leaf as the Sun's rays strike it. Secondly, the water vapour that is produced by evaporation passes out through holes called stomata in the underside of the leaf, and must be replaced. This loss of water causes the leaf to act like a pump and draw up water through the plant from the roots.

Water-conducting tissue

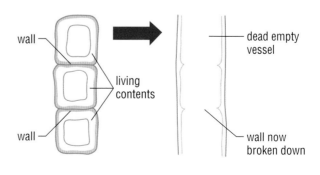

Figure 1.15 How a xylem vessel is made.

Water moves through the plant in xylem tissue. This tissue is made from xylem cells which die and form hollow tubes called xylem vessels (see Figure 1.14).

Water is needed in the leaf not only to keep the leaf cool, but also as a raw material for making food.

Food-conducting tissue

There is a second tissue which forms tubes – this is called phloem tissue. The tubes are made of living cells which transport the food made in the leaf to other parts of the plant.

38 Describe how water in the xylem tissue in a stem reaches the air outside the leaf.

From tissues to organs

Just as cells are arranged in an orderly way into tissues, tissues are arranged in an orderly way into groups called organs. An organ may perform one or more tasks to help keep an organism alive, and organs can contain different types of tissues.

The human stomach wall is an organ made from several tissues. These include muscle tissues which help churn up the food in the stomach, and tissues which form glands that produce food-digesting liquids.

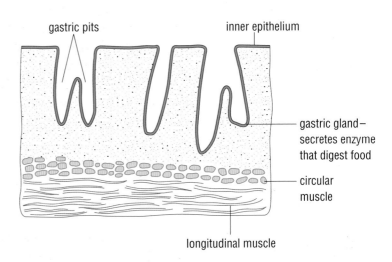

Figure 1.16   Section through stomach wall.

How new cells are made

Cell division

There are some organisms which have a body that is made from only one cell. *Amoeba* is an example. It lives on the bottom of ponds, and moves around by sending out projections from its sides. These projections are called pseudopodia. When a pseudopodium is formed the contents of the rest of the body flow into it. *Amoebae* breed by dividing in two as shown in Figure 1.17.

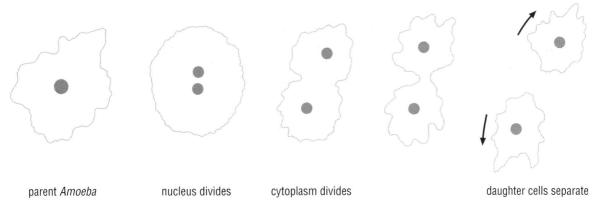

parent *Amoeba*   nucleus divides   cytoplasm divides   daughter cells separate

Figure 1.17   Stages in the division of an *Amoeba.*

39 Ink marks were placed at regular intervals along the root of a broad bean seedling but later the marks appeared as shown in Figure 1.18. What has happened?

mark the uppermost edge of the dish so that you know which is the top

broad beans clamped inside moist cotton wool

regular ink marks

start

two days later

Figure 1.18

Many cells are able to make new cells by dividing into two. This is necessary for organisms to grow and repair damaged tissue. They begin by making a copy of the nucleus, and then the cyptoplasm divides. For example, near the tip of the root are large numbers of cells which are produced by simple cell division. One of the pair of cells eventually grows longer while the other divides again.

Yeast is a fungus which has a body made from one cell. When the yeast cell divides, it produces a second smaller cell. The two cells do not separate at first, and

Figure 1.19   Yeast budding.

the smaller cell forms a structure called a bud. The cell in the bud may also divide so that each bud may produce buds of its own before all the cells separate.

Cells and reproduction

Most living things are produced by the joining of two special cells called gametes. Gametes are the reproductive cells of an organism.

In animals

In animals, the male produces a gamete called the sperm cell and the female produces a gamete called the egg cell (see Figure 1.12). These two gametes join together in a process called fertilisation (see Chapter 2).

In flowering plants

Plants also produce male and female gametes. The male gamete is a cell in the pollen grain and the female gamete is an egg cell in the ovule. Most flowers have both male and female parts as Figure 1.20 shows.

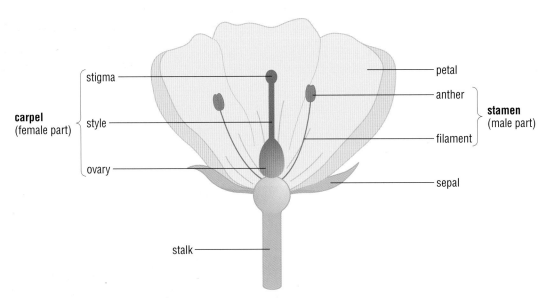

Figure 1.20 Parts of a typical insect-pollinated flower.

The outer part of the flower is formed by a ring of sepals called the calyx. Sepals are like small leaves and form a protective coat over the more delicate flower parts when the bud is developing.

Inside the sepals is a ring of petals called the corolla. The petals are usually large, brightly coloured parts of the flower.

Inside the ring of petals are the stamens which form the male part of the flower. Each stamen has two parts – the stalk, called the filament, and the pollen-producing organ, called the anther. The male gametes form inside the pollen grains.

Inside the ring of stamens is the female part of the flowers which is made up from one or more carpels which may group together to form a pistil. Each carpel has a pollen-receiving surface called a stigma. Beneath the stigma is the style. It is connected to the ovary which contains one or more ovules.

40 What are the differences between a sepal and a petal?

41 In what ways are stamens and carpels
  a) different and
  b) similar?

Pollen grains and pollination

When the pollen grains are fully formed in the anther it splits open to release them. Pollination occurs when pollen is transferred from an anther to a stigma. If the pollen goes from an anther to the stigma of the same flower or other flowers on the same plant the process is called self-pollination. Cross-pollination occurs if the pollen goes from the anther to the stigma of a flower on another plant of the same species. Most plants produce flowers that have both male and female reproductive parts. They avoid self-pollination in two ways. Firstly, the anther can release the pollen before the stigma is ready to receive it, or secondly,

Self-pollination

Cross-pollination

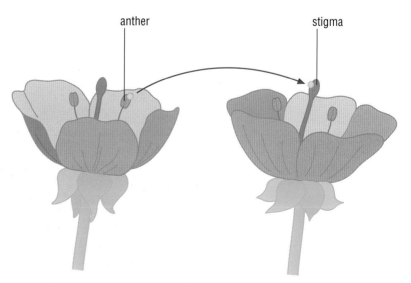

anther

stigma

Figure 1.21 Types of pollination in flowering plants.

the stigmas can be ready to receive pollen from other plants of the same species before their own anthers are ready to release their pollen.

There are two main ways in which the pollen grains are transferred from one flower to another for cross-pollination. They may be carried by insects or they may be carried by the wind. Pollen grains carried by insects may have a spiky surface which helps them stick to the hairs on the insect's body. Pollen grains carried by the wind are very small and light so that they can easily travel on air currents.

Insect- and wind-pollinated flowers

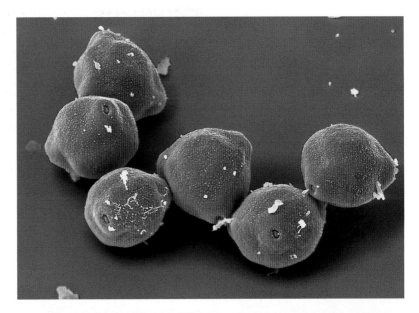

Figure 1.22   Pollen from a wind-pollinated plant (top) and from an insect-pollinated plant (bottom).

The flowers of insect-pollinated plants are different from the flowers of wind-pollinated plants.

Insect-pollinated flowers have a range of adaptations that attract insects. These adaptations include large colourful petals, scent and nectaries that produce a sugary liquid called nectar on which the insects feed. Some flowers produce more pollen than is needed for pollination and this may be taken as food by the pollinating insect. Many insect-pollinated plants, such as orchids, are adapted so that they attract just one species of insect. The shape and arrangement of the petals may allow one species of insect to enter a flower but keep out other species. The structure of the flowers encourages the transfer of pollen onto the insect and then onto the stigma of a plant of the same species. Short filaments keep the anthers inside the flower so that the insect can brush past them. The anthers of insect-pollinated flowers make a smaller amount of pollen than those of wind-pollinated flowers. Their stigma is often flat and held on a short style inside the flower so that the insect can easily land on it.

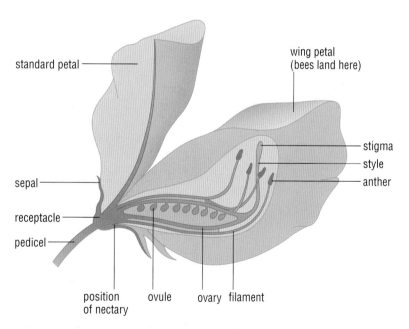

Figure 1.23   Insect-pollinated flower.

Wind-pollinated flowers are smaller than insect-pollinated flowers and do not show the adaptations shown above. They may have green petals and produce no nectar or scent. The flowers have long filaments which allow the anthers to sway outside the flower in the air currents. The anthers make a large amount of pollen and the stigma is a feathery structure which hangs outside the flower and forms a large surface area for catching pollen in the air.

42 Make a table to compare wind- and insect-pollinated plants.
43 Why does one method of pollination require much more pollen than the other method?
44 What is the difference between self-pollination and cross-pollination?

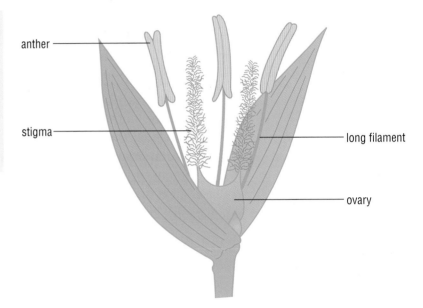

Figure 1.24 Wind-pollinated flower.

Fertilisation

After a pollen grain has reached the surface of a stigma it breaks open and forms a pollen tube. The male gamete that has travelled in the pollen grain moves down this tube. The pollen tube grows down through the stigma and the style into the ovary (see Figure 1.25 page 22). In the ovary are ovules, each containing a female gamete. When the tip of a pollen tube reaches an ovule the male gamete enters the ovule. It fuses with the female gamete in a process called fertilisation and a zygote is produced.

Bees and flowers

Karl von Frisch (1886–1982) was an Austrian zoologist who studied how bees communicate with each other. In 1973 he received a Nobel Prize for his work on animal behaviour.

The scent from a flower that has produced nectar travels through the air. It may stimulate the receptor cells of a honey bee and the insect flies towards it. As it gets closer, the bee also uses its eyes to find the flower. Its eyes are sensitive to ultraviolet light. This makes some of the pale markings we see in normal light stand out more distinctly to help the bee identify the flower. Some of the markings are lines running down the inside of the petal. They are called honey guides and direct the bee towards the nectar.

After landing on the flower the bee sticks its head between the stamens and probes the nectary with its mouth parts. While taking up the nectar it brushes past the anthers and pollen collects on the hairs of its back. When the bee has collected the nectar it flies on to the next flower and feeds again. Some of the pollen passes onto the stigma of the next flower.

The bee has stiff hairs on its front legs. Periodically it runs them through its body hair like a comb. This action collects the pollen off the bee's back and it is stored in structures, made from hairs on its back legs, called pollen baskets.

1 What attracts the bee to the flower? Which sense organs does it use?
2 How do you think that Karl von Frisch gathered information about the honey bee's behaviour?
3 How does the behaviour of the dancing bee help a colony of plants which have come into flower?
4 How do you think the hive of bees survive the winter when there are no flowers to feed on?
5 Why are hives of bees kept in orchards?

Figure A A bee in flight showing full pollen baskets.

When the bee swallows the nectar it collects in a cavity called the honey sac. The action of enzymes and the addition of other substances change the nectar into honey. After the bee has returned to its hive, it regurgitates the honey and passes it on to other bees working in the hive. They store it in the honeycomb. Also, the pollen is removed from the pollen baskets and stored.

The bee indicates the source of the nectar to the other bees in the hive by performing a dance on the honeycomb. The dance involves the bee moving in circles, waggling its abdomen and moving straight up and down on the vertical surface of the honeycomb. From this performance the other bees can tell the distance, direction and amount of nectar available and can set out to search for it.

For discussion
How useful are bees? Should we worry if there were fewer bees? Explain your conclusions.

pollen grain

stigma

style

course followed
by pollen tube

ovary wall

ovule

cavity

egg cell

egg cell
nucleus

pollen
tube
nucleus

45 What is the difference between pollination and fertilisation?

46 Trace the path of a male gamete nucleus from the time it forms in a pollen grain in an anther until the time it enters an ovule.

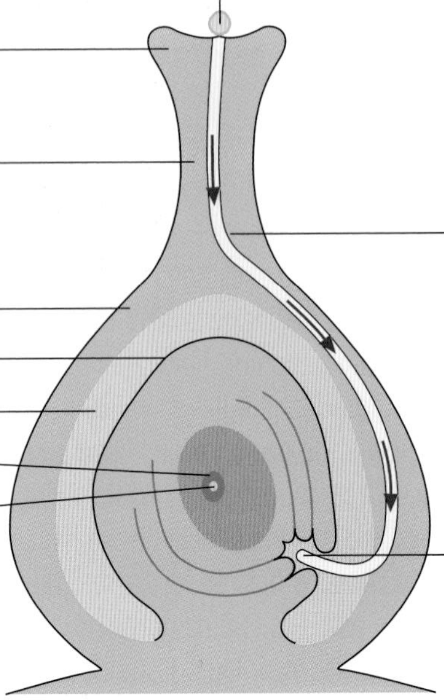

Figure 1.25 Fertilisation.

After fertilisation

The zygote undergoes repeated cell division to form the embryo plant. Structures that later become the root and shoot are developed and a food store is laid down. While these changes are taking place inside the ovule the outer part of the ovule is forming a tough coat. When the changes are complete the ovule has become a seed (see Figure 1.26). As the seeds are forming other changes are taking place. The petals and stamens fall away. The sepals usually fall away too but sometimes, as in the tomato plant, they may stay in place. The stigma and style wither and the ovary changes into a fruit.

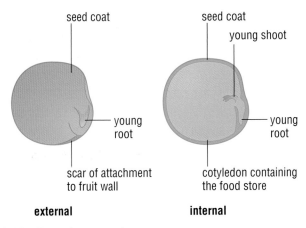

seed coat

seed coat

young shoot

young
root

young
root

scar of attachment
to fruit wall

cotyledon containing
the food store

external

internal

Figure 1.26 Parts of a pea seed.

Types of fruit

A fruit forms from the parts of the flower that continue to grow after fertilisation. There are two main types of fruit – dry fruits and succulent fruits.

Dry fruits have a wide variety of forms. They may form pods, such as those holding peas and beans, they may be woody nuts, such as acorns or hazelnuts, or grains like the fruits of wheat, oats and grasses.

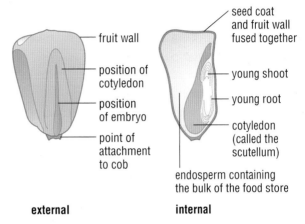

Figure 1.27   Parts of a cereal grain.

Succulent fruits have a soft fleshy part. They may have a seed inside a woody skin which forms a 'stone' in the fruit, as in the cherry and peach. Many succulent fruits do not have a stone but contain a large number of smaller seeds, as in the tomato and orange.

Some fruits, such as apples, are called false fruits because their fleshy part does not grow from part of the flower but from the receptacle on which the flower grows.

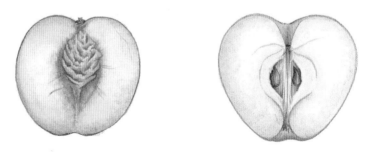

Figure 1.28   A peach (succulent fruit) and an apple (false fruit).

◆ SUMMARY ◆

◆ The part of a body that performs a task to keep a living thing alive is called an organ (*see page 1*).

◆ There are ten organ systems in the human body. They are the sensory system, the nervous system, the respiratory system, the digestive system, the circulatory system, the excretory system, the skeletal system, the muscle system, the endocrine system and the reproductive system (*see pages 1–2*).

◆ There are four main organs in the body of a flowering plant. They are the root, stem, flower and leaf (*see page 3*).

◆ The microscope is used to observe very small living things or the cells of larger living things (*see page 5*).

◆ There are special techniques for finding the size of a small object with a microscope (*see pages 6–7*).

◆ The bodies of plants and animals are made of cells. The basic parts of the cell are the nucleus, cytoplasm and cell membrane (*see page 8*).

◆ In a plant cell there is a cellulose cell wall and a vacuole (*see page 9*).

◆ Cells have different forms for different functions. They are adapted to perform specific tasks in the body and life of the organism (*see pages 10–12*).

◆ Cells form groups called tissues (*see pages 12–14*).

◆ New cells are made when a cell divides (*see pages 14–16*).

◆ Plants and animals produce sex cells called male and female gametes (*see page 16*).

◆ The parts of a flower are adapted for reproduction (*see page 16*).

◆ Pollen is transferred in pollination (*see page 17*).

◆ Wind and insect pollination are the two main kinds of pollination (*see pages 18–20*).

◆ When fertilisation takes place in flowering plants, seeds and fruits are produced (*see pages 20, 22, 23*).

End of chapter questions

1 If organisms as complex as us have developed on a planet similar to Earth but with a stronger force of gravity, what body features would you expect them to have?

2 Pollen grains will grow pollen tubes if they are placed in a sugar solution of a certain concentration. The results in the tables reflect results sometimes produced in investigations.

An experiment was set up to find out the concentration of sugar that would cause the pollen grains of a plant to produce pollen tubes. Here is a table of the results.

Table 1.1

Concentration of sugar solution/%	5	10	15	20
Number of pollen grains	20	20	20	20
Number of grains with tubes	4	18	3	0

a) What percentage of pollen grains produced tubes in each solution?

b) What would you expect the concentration of sugar in the stigmas of the flowers to be?

When the pollen of a second type of plant was investigated the following results were obtained.

Table 1.2

Concentration of sugar solution/%	5	10	15	20
Number of pollen grains	20	20	20	20
Number of grains with tubes	0	1	3	8

c) Why was it decided to take the investigation further?

d) What do you think was done to take the investigation further?

2 Reproduction

All animals follow a particular pattern of development. Like plants, they grow from a fertilised egg and after hatching or being born they grow until they are fully mature. At this stage they are capable of reproduction. In almost all kinds of animal there are those individuals that produce eggs – the females, and those that produce sperm – the males. For reproduction to occur the sperm from the male has to reach the eggs of the female, and the head of one sperm has to enter each egg. When the sperm and egg nuclei fuse, fertilisation has occurred.

Fertilisation

There are two kinds of fertilisation – external fertilisation and internal fertilisation.

External fertilisation

This process occurs in most fish and in many amphibians. The eggs and sperm are shed into the water around the animals and fertilisation takes place there. When fish are about to reproduce, the males and females swim close together. This increases the chance of the sperm cells mixing with the eggs when both are shed into the water. In amphibians such as the common frog, the male climbs on the female's back and stays there until the eggs and sperm have both been released. This behaviour increases the chances of the eggs and sperm meeting, and of external fertilisation being successful.

Female animals which have their eggs fertilised externally produce large numbers of eggs. The female common frog, for example, lays up to two thousand eggs. By laying large numbers of eggs, the female increases the chances of fertilisation. This is necessary because even when the parents are close together, some of the eggs may drift away and some sperm may swim in the wrong direction. If only a small number of eggs were produced, the chances of an egg meeting a sperm would be much reduced.

Internal fertilisation

In this process the eggs remain inside the female's body and the sperm are placed inside her body by the male. Internal fertilisation occurs in insects, reptiles, birds and mammals. In insects, reptiles and mammals the males are equipped with a tube which delivers the sperm into the female's body. In mammals this tube is called a

◆ SUMMARY ◆

◆ All animals follow a particular pattern of development (*see page 26*)
◆ There are two kinds of fertilisation – external fertilisation and internal fertilisation (*see page 26*)
◆ Animals may or may not show any parental care (*see page 27*)
◆ Changes in the human body at puberty are brought about by the sex hormones (*see page 32*)
◆ The male and female reproductive organs have differences and similarities (*see pages 35–36*)
◆ The menstrual cycle occurs due to the monthly release of an egg (*see pages 36–37*)
◆ Fertilisation is the fusion of a sperm nucleus with an egg nucleus (*see page 40*)
◆ Development of the embryo takes place in the uterus (*see pages 42–43*)
◆ Twins may be identical or non-identical (*see page 42*).
◆ The placenta and the amnion play important parts in the development of the embryo and fetus (*see pages 43–45*).
◆ The fetus may be lost by miscarriage or abortion (*see page 45*).
◆ The uterus and abdominal muscles are used in the birth process (*see page 46*).
◆ Many body changes take place in the child's early life (*see pages 47–48*).

End of chapter questions

1 How does a knowledge of the basic facts of reproduction help someone going through puberty?
2 Gillian and Barry have got a record of their heights (in cm), measured once a year from birth until they were 17 years old.

Gillian's record is: 51, 75, 87, 95, 102, 108, 115, 122, 127, 132, 138, 144, 151, 157, 159, 161, 162, 163

Barry's record is: 51, 76, 88, 96, 103, 109, 117, 123, 129, 135, 139, 143, 149, 154, 160, 167, 173, 174

a) Make a table of the results.
b) Plot each set of results on the same graph.
c) At what ages were Gillian and Barry the same height?
d) Who grew more quickly between the ages of
  i) 8 and 9,
  ii) 11 and 13,
  iii) 14 and 17?

For discussion
'People should develop a responsible attitude towards sex.' What do you think this means?

3 Survival

If you look out across the countryside you may see fields, hedges, woods, ponds and maybe a river. Most of the living things you see will be plants ranging in size from green slime on rocks to the tallest tree in a wood. You may see some birds flying across the countryside and a few insects moving through the air around you. There may be a slug slowly moving across your path or a squirrel scampering away through the branches of a nearby tree. The scene may look too complicated to investigate scientifically, but the study of ecology was established at the beginning of the 20th Century to do just this.

Ecology means the study of living things and where they live. The home area of a living thing is called its habitat. Two examples of habitats are a wood and a pond. The country scene in Figure 3.1 can be divided into a number of different habitats for further investigation.

1 What habitats can you see in Figure 3.1?

Figure 3.1 A countryside scene from a hilltop showing a range of habitats.

A living thing in its habitat is affected by two different kinds of factors. They are abiotic factors and biotic factors. Abiotic factors are not due to living things and include temperature, wind strength, amount of light and moisture. Biotic factors are due to living things and include plants and animals as sources of food, other organisms competing for space, and predators.

Adaptation

A living thing survives in its habitat because it can cope with the abiotic and biotic factors. It copes if its body is adapted to the conditions of the habitat. Changes that have taken place in the structures of different species over time that help them survive are called adaptations.

Some adaptations in plants

Grass

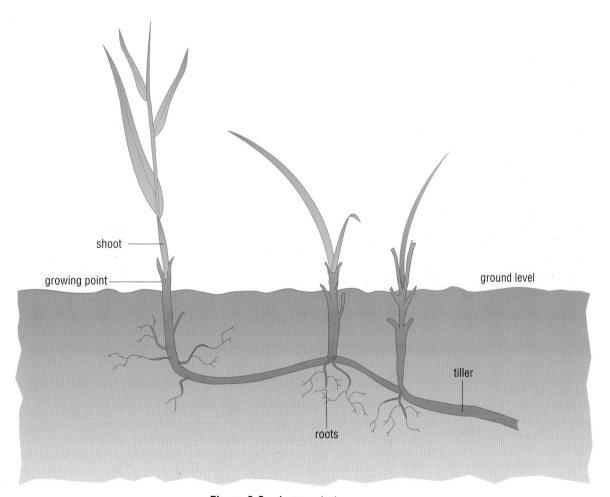

Figure 3.2 A grass plant.

One of the commonest land plants is grass. Unlike many plants it survives in areas where there are grazing animals. Grass has growing points that are below the reach of the grazers' mouths. This means that when most of the leaves of the grass are removed and eaten, the growing points can produce new leaves. The grazing animals also eat the grass flowers, but grass can also reproduce asexually by sending out side-shoots, called tillers, that grow along or just below the soil surface. Buds along the tiller produce new grass plants. The colony produced in this way binds with the soil to form turf which is hard wearing and is not destroyed by the feet of the grazing animals. These adaptations allow the grass to survive.

2 Why do the low growing points of grasses help them to survive?

Water plants

The roots of land plants have oxygen around them in the air spaces in the soil. In the waterlogged mud at the bottom of a pond there is very little oxygen for the root cells. The stems of water plants have cavities in them through which air can pass to the roots in order to overcome this problem.

3 In what ways are the features of a plant living in water different from a plant living on land?

Figure 3.3 A pond with a range of water plants.

Water plants use the gases they produce to hold their bodies up in the water and therefore do not need strong, supporting tissues like land plants. Minerals can be taken in from the water through the shoot surfaces of the water plant, leaving the root to act as an anchor. The leaves of submerged water plants are thin, allowing minerals in the water to pass into them easily. The leaves also have feathery structures that make a large surface area in contact with the water. This further helps the plant to take in all the essential minerals.

Floating water plants like duckweed have a root that acts as a stabiliser.

Daily adaptations in plants

Flowers

Some plants such as the crocus and the tulip open their flowers during the day and close them at night. The flowers are open in the day so that insects may visit them for nectar, and in return transport pollen from one flower to another to bring about pollination. The flowers close at night to protect the delicate structures inside the petals from low temperatures and from dew. The dew could wash the pollen off the stamens (see page 16) so that it cannot be picked up and transported by the insects. The night scented stock is an unusual plant in that its flowers open in the evening and close during the day. This adaptation allows moths to visit the flowers and pollinate them.

Leaves

A few plants make movements of their leaves over a twenty-four hour period. Leaves not only make food but they also provide a large surface for the evaporation of water, which in turn helps to draw water through the plant from the roots. In clover, for example, each leaf is divided into three leaflets. During the day the leaflets spread out and become horizontal. In this position they are best placed for receiving sunlight to make food, and to lose water to the air which causes the plant to draw up water from the roots. The water is needed to make food, and it also helps to keep the leaves cool in the sunlight. In the evening the leaflets fold close together. This helps them to lose less water when the plant is not making food.

The wood sorrel is another plant which moves its leaves over a twenty-four hour period. At night it lets its leaves droop to conserve water. Plants can make those movements by moving water about inside their bodies. When water is concentrated in one place it makes the cells swell up. When part of it is removed from a place the cells sag. This swelling and sagging of the cells allows the plants to move slowly.

4 Do the petals of a crocus and the leaves of a clover plant move for the same reason? Explain your answer.

5 In what ways are plants adapted to survive winter conditions?

Seasonal adaptations in plants

The abiotic factors in a habitat change with the seasons. The grass plant is adapted to survive winter conditions but its short roots make it dependent upon the upper regions of the soil staying damp. In drought conditions the soil dries out and the grass dies. Daffodils are adapted to winter conditions as the leaves above the ground die and the plant forms a bulb in the soil. Bark is an adaptation of trees that provides a protective insulating layer around the woody shoot in winter.

Plants that float on the open water of a pond in spring and summer do not remain there in the winter. Duckweed produces individuals that sink to the pond floor; the water plant called frogbit produces heavy seeds. The plants around the water's edge die back and survive in the mud as thick stems called rhizomes.

Some adaptations in animals

Land animal

The tawny owl has several adaptations that allow it to catch mice at night. It has large eyes that are sensitive to the low intensity of light in the countryside at night. These allow it to see to fly safely. The edges of some of the owl's wing feathers are shaped to move noiselessly through the air when the bird beats its wings. This prevents the mouse's keen sense of hearing from detecting the owl approaching in flight. The owl has sharp talons on its toes that act as daggers, to kill its prey quickly and to help carry the prey away to be eaten at a safe perch.

6 What adaptations does the tawny owl have that allow it to detect its prey, approach its prey and attack its prey?

7 Why should the owl kill its prey quickly?

8 What adaptations do you think a mouse may have to help it survive a predator's attack?

Figure 3.4   A tawny owl, swooping down on a wood mouse.

Water animal

Although it lives underwater the diving beetle breathes air. It comes to the surface and pushes the tip of its abdomen out of the water. The beetle raises its wing covers and takes in air through breathing holes, called spiracles, on its back. (In insects living on land the spiracles are on the side of the body.) When the beetle lowers its wing covers more air is trapped in the hairs between them. It is able to breathe this air while it swims underwater. Diving beetles feed on a range of foods, including small fish, tadpoles and other insects.

Figure 3.5 A diving beetle feeding on an earthworm.

Daily adaptation in animals

Animals are adapted to being active at certain times of the day and resting at other times. At night most birds roost (sleep) but as soon as it is light they may start flying about in search of food. They have large eyes and consequently have excellent vision, which is essential for them to fly, land and search for food.

At night they are replaced in the air by bats. These animals roost in the day and come out at dusk to hunt for flying insects. Bats do not use their eyes but have developed an echo-location system. They send out very high-pitched squeaks that we cannot hear. These sounds reflect off all the surfaces around the bat and travel back to the bat's ears. The bat uses the information from these sounds to work out the distance, size and shape of objects around it. This allows the bat to fly safely and detect insects in the air, which it can swoop down and eat.

Many insects such as butterflies, bees and wasps are active and fly during the day. At night moths take to the air to search for food.

The squirrel is a mammal that is active during the day. Deer may also be active but they hide away in vegetation. Field mice and voles may be active during the day but hide in the grass and other low vegetation. These animals are also active for periods at night, when they are at risk of falling prey to night predators such as the fox and the owl.

While darkness is the major feature of a habitat at night there is also a second feature – humidity. At night all surfaces in the habitat cool down and the air cools down too. This causes water vapour in the air to condense and form dew. The increase in humidity is ideal for animals such as slugs and woodlice, which have difficulty retaining water in their bodies in dry conditions. They hide away in damp places during the day but as more places in the habitat become damp at night they become more active and roam freely searching for food. In the morning as the humidity decreases they hide away again somewhere damp.

9 Imagine that you are camping in a wood.
   a) What animals may you expect to see during the day?
   b) What animals are active in the evening?

Seasonal adaptations in animals

The roe deer (see Figure 3.6) lives in woodlands. In the spring and summer when the weather is warm it has a coat of short hair to keep it cool. In the autumn and winter it grows longer hair that traps an insulating layer of air next to its skin. This reduces the loss of heat from its body.

The stoat grows a white coat in the winter which loses less heat than its darker summer coat. The stoat preys on rabbits and its white coat may also give it some camouflage when the countryside is covered in snow.

The ptarmigan is about the size of a hen. It lives in the north of Scotland, northern Europe and Canada. In summer it has a brown plumage that helps it hide away from predators while it nests and rears its young. In winter it has a white plumage that reduces the heat lost from its body and gives it camouflage. Feathers grow over its toes and make its feet into snowshoes which allow it to walk across the snow without sinking.

10 How do the adaptations of
   a) the roe deer,
   b) the stoat and
   c) the ptarmigan help them survive in the winter?
11 How might their winter adaptations affect their lives if they kept them through the spring and summer?

Figure 3.6   A roe deer in summer (left) and a ptarmigan in winter (right).

Adaptations for feeding

There are two main ways in which animals are adapted for feeding. There are animals which are adapted for feeding on plant foods (these animals are called herbivores), and those that are adapted for feeding on other animals (carnivores).

Herbivores

When people think of herbivores they tend to think of herbivorous mammals such as the rabbit or the deer. Herbivores exist in other animal groups too. For example caterpillars, which are insects, and slugs and snails, which are molluscs, are all herbivores. Plant material is tough so herbivorous animals have adaptations that allow them to break it up for digestion. Herbivorous mammals such as the cow and sheep have large, strong back teeth that they use for grinding up the food. Caterpillars have strong jaws for nibbling along the edge of a leaf, while slugs and snails have a tongue covered in tiny teeth which they use like sandpaper to rasp away at the surface of their food.

Herbivorous animals are the prey of carnivorous animals, so they have developed features which help them reduce their chances of being caught and eaten. Many herbivores, from caterpillars to giraffes, have body colours which help them blend into their surroundings –

they have camouflage. Some herbivores, such as deer (see Figure 3.6 page 57), may also hide away during the day in vegetation and come out into the open at night when it is difficult for carnivores to see them.

Herbivorous mammals such as the rabbit (see Figure 4.2 page 66) have eyes on the sides of their head. This gives them a very wide field of vision, enabling them to see a carnivore approaching. Rabbits, like many herbivorous animals, have large ears which can be turned to face almost every direction so that the sound of an approaching predator can be detected.

Carnivores

Just as people think of a rabbit as a herbivore they may think that all carnivores are like the fox or wolf. Carnivores, like herbivores, come in all shapes and sizes. Spiders, for example, are carnivores. They set web traps to catch their prey. The frog is also a carnivore, and can flick out its tongue very quickly to catch flies. Most carnivorous mammals have large conical canine teeth (see page 90) for stabbing their prey, and molars which are adapted for holding bones while the jaw muscles press on them to crack them open for their marrow. The shrew belongs to a group of mammals which feed mainly on insects, called the insectivores (see Table 4.3 page 72). Its teeth are pointed, making the jaws look like those of a miniature crocodile. This arrangement of teeth allows the shrew to catch hold of the tough body of an insect and chew it up.

Animals which catch prey are called predators. Predatory birds such as eagles, hawks and owls are known as birds of prey and are adapted for catching and feeding on other animals. They have long claws on their feet called talons, which they use to grab and stab their prey. They also have hooked beaks for ripping up their prey into smaller pieces that are easy to swallow. Carnivorous birds and mammals share an adaptation. They both have eyes that face forwards. This means that the field of view of each eye partly overlaps the field of view of the other eye, and this allows the animal to judge distance. Without this overlap judging distance can be very difficult. You can test this yourself by putting a pen and its top on the table. Close one eye, look at the two objects then pick them up and try quickly to put the top on the pen. The chances are that the first time you try this you will miss. Carnivorous animals need to be

12 a) Design a bark-feeding mammal that burrows its way from tree to tree.
b) What adaptations would you give it to protect it from predators?

13 Design a bird that feeds on the mammal you invented for Question 12. Explain the reasons for the features you give it.
14 A starling pulls up a worm and eats it. Later a sparrow hawk attacks and kills the starling and carries it away for a meal.
Do these observations support the idea that carnivores are always predators? Explain your answer.

able to judge distance extremely accurately to pounce on their prey. If they miss, they go hungry.

As herbivorous mammals are constantly looking, listening and even sniffing the air for signs of an approaching predator, predators themselves have to take care when they are hunting. Some predators such as lions even set up an ambush to catch their prey.

Special relationships

A predator is a carnivore and the animal that it feeds on is its prey. A predator may have a wide range of prey. A weasel, for example, feeds on frogs, mice, voles, rats, small birds and moles. In the prey–predator relationship the predator survives by killing the prey. Young, old or sickly prey are the easiest animals to catch. The prey may sometimes try to avoid being caught by having camouflage or being able to move fast.

A parasite lives on or in another organism and feeds on it. The organism that contains the parasite is called the host. A head louse is an ectoparasite. It lives on the outside of the body and feeds on blood by piercing the skin on the scalp. The tapeworm is an endoparasite. It lives in the small intestine and feeds on the digested food. In the parasite–host relationship both organisms stay alive but the host is harmed (sometimes fatally) by the presence of the parasite. Some fungi are parasites on green plants and frequently kill them. Mistletoe is a green plant that is a semi-parasite. It has green leaves and can photosynthesise but it takes its minerals from the apple or poplar tree on which it is growing.

There are some organisms that are able to live apart but benefit when they live together. This relationship is called commensalism. Feeding may be only a part of this relationship. For example, the hermit crab often has sea anemones on its shell. They provide it with some camouflage and the messy feeding habits of the crab make a cloud of food particles in the water that the sea anemones can feed on.

Mutualism is a relationship where both organisms need each other to survive. The termite feeds on wood, which is made of cellulose. Protoctista that make an enzyme to digest cellulose live in the termite's gut. They allow the termite to digest the wood and in return the termite provides them with a home. The lichen is another example of mutualism. It consists of a fungus and an alga growing together. The fungus provides the support in which the alga can grow. The alga makes food by photosynthesis using the water that the fungus has stored. The fungus also takes in minerals to be converted into materials for growth. Together the two organisms form a structure that can live on the surfaces of rocks in harsh conditions where other organisms cannot survive.

1 Write down six examples of predators and their prey. For each pair say how each animal is adapted to catch prey or avoid being eaten.
2 Why is it a disadvantage if the parasite kills the host?
3 Why is mistletoe called a semi-parasite?
4 What is the difference between commensalism and mutualism?
5 How might a predator help the population of its prey?

Figure A Head louse.

Figure B Hermit crab with sea anemones on its shell.

Looking for links between animals and plants

All the individuals of a species in a particular habitat make up a group called a population. All the populations of the different species in the habitat make up a group called a community. Ecologists look for links between the animals and plants in the community to understand how they live together.

Most animals spend a large part of their time searching for food. By observing animals in their habitat the food of each species can be identified.

Living things can be grouped according to how they feed. Plants make their own food by photosynthesis and by taking in minerals. They are called the producers of food. Animals that feed on plants are called primary consumers. They are also known as herbivores. Animals that feed on primary consumers are called secondary consumers. They are also known as carnivores. The highest level consumer in a food chain is called the top carnivore. Some animals such as bears feed on both plants and animals. They are called omnivores. An omnivore feeds as a primary consumer when it feeds on plants, and as a secondary or higher level consumer when it feeds on animals.

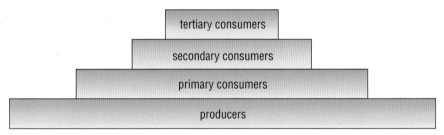

Figure 3.7 This diagram shows how the number of producers and consumers compare in a habitat. The bigger the block, the bigger the number of organisms.

Food chains

The information about how food passes from one species to another in a habitat is set out as a food chain. For example, it may be discovered that a plant is eaten by a beetle which in turn is eaten by a shrew and that the shrew is eaten by an owl. This information can be shown as follows:

plant → beetle → shrew → owl

Once a food chain has been worked out further studies can be done on the species in it.

15 In the food chain identify
   a) the primary consumer,
   b) the secondary consumer and
   c) the tertiary consumer.
16 In the food chain identify the herbivore and the carnivores.

17 Construct some food chains with humans in them.

18 In the food chains you have constructed are humans classed as herbivores, carnivores or omnivores?

Energy and the food chain

While animals need food substances from their meals to nourish their bodies they also need energy to keep their bodies alive. The source of energy for almost all food chains is the Sun (for more details on the energy from the Sun see *Physics Now! 11–14* second edition). Some of the energy in sunlight is trapped in the leaves of plants and stored in food that the plant makes. When a herbivore eats the plant the energy is transferred to the body of the herbivore. Some of this energy is used up by the herbivore to keep it alive, but some is stored in its tissues. When a carnivore eats a herbivore it takes in the energy from the herbivore's body and uses it to keep itself alive. The food chain is an example of an energy transfer system.

A few food chains do not have the Sun as a source of energy. They are found around hot springs deep in the ocean. Bacteria use hydrogen sulphide in the water as a source of energy, and crabs and worms feed on the bacteria.

Food webs

When several food chains are studied in the habitat some species may appear in more than one. For example, a badger eats blackberries and also eats snails. The two food chains it appears in are:

blackberry → badger
plant → snail → badger

The two food chains can then be linked together as:

blackberry → badger
↗
plant → snail

When all the food chains in the habitat are linked up they make a food web.

A food web shows the movement of food through a habitat. It can also be used to help predict what might happen if one of the links in a food web was absent. Look at Figure 3.8 overleaf and think of each animal shown there not just as one animal but as the whole population of that species of animal in the wood. If you think of each animal in the plural, such as voles and finches, it may help you think about animal populations.

19 Is a badger a herbivore, a carnivore or an omnivore? Explain your answer.

20 How may the numbers of other species in the wood change if each of the following was removed in turn:
a) fox
b) seeds?

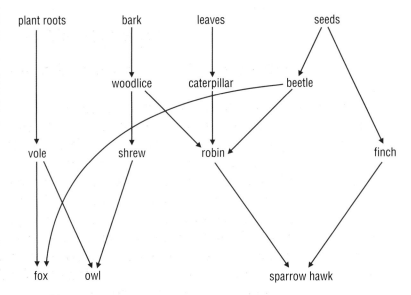

Figure 3.8   Examples of a food web in a woodland habitat.

Now imagine that the trees had a disease that made their leaves fall off. The caterpillars would starve and die, and they would not be available as food for the robins. This means that the robins must eat more beetles and woodlice if they are not to go hungry. The reduction in the number of woodlice would affect the shrews, as this is their only food. The beetle population would also fall, forcing the fox to search out more voles to eat.

◆ SUMMARY ◆

- There are two different kinds of factors in a habitat – abiotic and biotic factors *(see page 51)*.
- Plants are adapted to their habitats *(see page 51)*.
- Some plants are adapted to change during the day *(see page 53)*.
- Plants are adapted to changes that occur with the seasons *(see page 54)*.
- Animals are adapted to their habitats *(see page 54)*.
- Animals are adapted to changes that occur during the day *(see pages 55–56)*.
- Animals are adapted to changes that occur with the seasons *(see page 56)*.
- Some animals are adapted for feeding on plants *(see page 57)*.
- Some animals are adapted for feeding on other animals *(see page 58)*.
- Plants and animals are linked to each other by the way animals feed *(see page 60)*.
- Food passes from one species to another along the food chain *(see page 60)*.
- A food chain is an energy transfer system *(see page 61)*.
- Food chains link together to form food webs *(see pages 61–62)*.

End of chapter questions

1 Construct a food web for the African plains from the following information.

Giraffes feed on trees, elephants feed on trees, eland feed on trees and bushes, hunting dogs feed on eland and zebra, finches feed on bushes, mice feed on roots, baboons feed on roots and locusts, gazelles, zebras and locusts feed on grass, foxes feed on mice, lions feed on eland, zebra and gazelle, hawks feed on finches, eagles feed on baboons, foxes and gazelles.

You may like to write the name of each living thing on a separate card and arrange the cards on a sheet of paper. You could write arrows on the paper between the cards but use pencil at first as you may find that you have to move the cards about to make the food web tidy.

2 Use the food web you have made for Question 1 to answer these questions.
- **a)** Which living organisms are producers?
- **b)** Which animals are herbivores?
- **c)** Which animals are carnivores?
- **d)** Which animal is an omnivore?

4 Looking at living things

You are probably aware that there are large numbers of different living things. Just think of the animals that you may see at a zoo, or the different plants that you may see in a park. You may also be aware that even among the same kind of living things there is variety. Just think of how you recognise different people by their own characteristic features. When scientists are looking at the features of living things they make careful observations and accurate records of what they see.

1 How are the leaves arranged in plant A and plant B in Figure 4.1?
2 How are the flowers arranged in A and B?

Making observations

When you make observations you look closely and with a purpose. For example, you may *look* at a plant and just see its flowers and leaves, but if you *observed* a plant you could study it to find out how the leaves and flowers are arranged on the stem. Leaves can be arranged in many ways, for example, they may grow alternately along a stem or they may be arranged in pairs. Flowers may be arranged singly or in columns.

B   St John's wort

A   Rosebay willow herb

Figure 4.1   The leaves and flowers in two plants.

Drawing specimens

Explorers of the 17th and 18th Centuries collected specimens of the plants and animals they found and brought them back to the scientists in Europe for further study. Many of the living things died during the journey and by the time they arrived their remains were decayed and of little use to the scientists. Even when the specimens were kept in a preservative their colours would be lost or some other feature would change. To solve the problem of showing how these living things appeared in their habitats, artists accompanied the explorers and drew pictures of the plants and animals that were discovered. The scientists back in Europe could then use both the specimens and the pictures to help them study and classify the new living things that were being discovered.

1 Why were artists taken on explorations in the 17th and 18th Centuries?
2 Why do you think artists are used much less in expeditions today?
3 Why might an organism be drawn ×5? Give an example. Why might another organism be drawn ×$\frac{1}{2}$? Give an example.
4 What is the true size of the living things in these drawings?

Figure A   17th Century biological drawing.

Biological drawings of specimens are still made today. The size of the specimen is usually indicated in one of two ways. A line may be drawn next to the picture to indicate the length of the specimen, or the drawing may have ×5 or ×$\frac{1}{2}$ next to it. The × symbol means times larger or smaller; the number gives an indication of the size. For example, ×5 means the drawing is five times larger than the specimen, and ×$\frac{1}{2}$ means the drawing is half the size of the specimen.

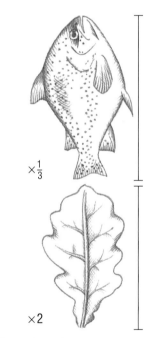

Figure B   A leaf and a fish.

Variation between species

Many living things have certain features in common. For example a cat, a monkey and a rabbit have ears and a tail. However, these features vary from one kind of animal to the next. In the species shown in the photographs on page 66, the external parts of the ears of the rabbit are longer than the ears of the cat. The external parts of the monkey's ears are on the side of its head while the other two animals have them on the top. The cat and the monkey have long tails but the monkey's tail is prehensile, which means it can wrap it around a branch for support while it hangs from a tree to collect fruit. (Only monkeys that come from South America have prehensile tails.) A rabbit's tail is much shorter than the cat's tail and the monkey's tail. These variations in features are used to separate living things into groups and form a classification system which is used worldwide.

Figure 4.2 A cat, a rabbit and a South American monkey.

Variation within a species

The individuals in a species are not identical. Each one differs from all the others in many small ways. For example, one person may have dark hair, blue eyes and ears with lobes while another person may have fair hair, brown eyes and ears without lobes. Another person may have different combinations of these features.

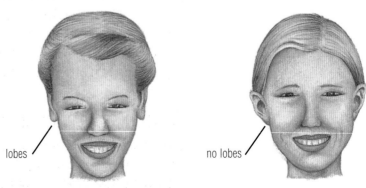

lobes   no lobes

Figure 4.3 Ears with and without ear lobes.

There are two kinds of variation that occur in a species. They are continuous variation and discontinuous variation.

Continuous variation

A feature that shows continuous variation may vary in only a small amount from one individual to the next, but when the variations of a number of individuals are compared they form a wide range. Examples include the range of values seen in different heights or body masses.

Discontinuous variation

A feature that shows discontinuous variation shows a small number of distinct conditions, such as being male or female and having ear lobes or no ear lobes. There is not a range of stages between the two as there is between a short person and a tall person. However, there are very few examples of discontinuous variation in humans.

Figure 4.4   Members of a family.

For discussion
Look at this photograph of a family. What features do the members of the family have in common? Which features are found in more than one generation?

The causes of variation

Some members of the family in Figure 4.4 have similar features. They are found in different generations which suggests that the features could be inherited. In fact we inherit many features, and the way they are passed on is explained in Chapter 9. Some variations may also be due to the environment.

Variation and the environment

The environment can affect the features of a living organism. For example, if some seedlings of a plant are grown in the dark and some in the light they will have

3 How else could the environment affect the development of an organism? Give another example for a plant and an animal.

different features. Those grown in the dark will be tall, spindly and have yellow leaves, while those grown in the light will have shorter, firmer stems with larger leaves that are green. Lack of food in the environment makes animals become thin. It can also slow down the growth of young animals.

Signs of life

Before something is sorted into a group of living things it must meet certain criteria to show that it is alive. A living thing must have seven special features. These are called the characteristics of life. The characteristics are feeding, respiring, moving, growing, excreting (getting rid of waste), breeding, and irritability (being sensitive to the surroundings).

Animal life

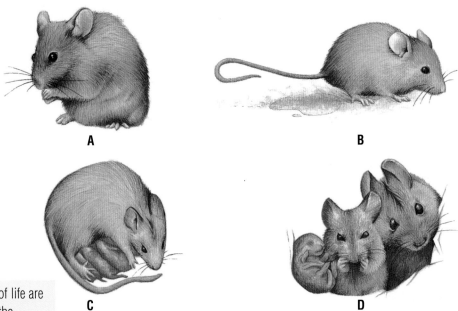

A

B

C

D

4 Which characteristics of life are shown by the mice in the pictures A–D in Figure 4.5?

Figure 4.5   Four of the characteristics of life.

Plant life

Green plants make food from oxygen in the air and water by using energy from sunlight. Chemicals in the soil are also needed, but in very small amounts. All plant cells respire and gaseous exchange takes place through their leaves. Plants move as they grow and can spread out over the ground (see page 150). Wastes may also be stored in the leaves. Green plants are sensitive to light and grow towards it. Plants reproduce by making seeds or spores. Some plants reproduce by making copies of themselves.

Figure 4.6   Seedlings growing towards the light.

5 How is a green plant's way of feeding different from an animal's way of feeding?

For discussion

A car can have five characteristics of life. What are they and how does the car show them?

If there are drought conditions, why might a plant produce seeds rather than grow new plantlets?

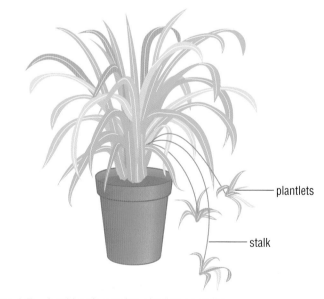

plantlets

stalk

Figure 4.7   A spider plant makes plantlets on stalks.

Life on Mars?

In 1976 two Viking spacecraft reached Mars. When they arrived each spacecraft split in two. One part, called the Orbiter, travelled around the planet taking photographs and measuring temperatures. The second part, called the Lander, touched down on the planet's surface. Although the Lander measured the Martian weather and had equipment on board to detect vibrations like earthquakes, the experiments that everyone wanted to know about were the ones to test the soil for signs of life. There was great excitement when one of the soil samples began to froth. It could have been caused by something respiring – a sign of life. After a few weeks of further investigation it was decided that the froth was most probably made by a chemical reaction not connected to a living organism.

There are signs that water once ran over the planet's surface, so some scientists believe that there could still be life there, but that it is buried deeper in the soil than the Lander could reach.

In 1997 the Pathfinder probe reached Mars and a vehicle called the Sojourner travelled around the landing site to take photographs and perform tests on rocks. It did not have any experiments on board to test for life.

Further observations on the plane have shown that some water is present. As living things need water there is a chance that when the water is examined some life forms may be found.

1 What were the two parts of the Viking spacecraft and what did each part do?
2 Why could the froth be a sign that something was respiring?
3 What are the advantages and disadvantages of bringing back Martian rock to Earth to look for organisms?
4 Can any of the disadvantages be overcome? Explain your answer.

Classifying living things

Living things are put into groups so that they can be studied more easily. The largest groups are called kingdoms. Scientists have now named five kingdoms. Each kingdom contains a large number of living things that all have a few major features in common. Table 4.1 shows the features that are used to place living things in either the animal or plant kingdom.

Table 4.1 The features of living things in the plant and animal kingdoms.

Plant kingdom	Animal kingdom
Make their own food from air, water, sunlight and chemicals in the soil	Cannot make their own food. Eat plants and animals
Body contains cellulose for support	Body does not contain cellulose
Have the green pigment chlorophyll	Do not have chlorophyll
Stay in one place	Move about

The way that the animal kingdom is divided up into subgroups is described on the following pages. A similar way of subgrouping is used to divide up the other kingdoms.

Subgroups of the animal kingdom

Each kingdom is divided into groups called phyla (singular: phylum). Each phylum contains living things with more similarities. For example, in the phylum Arthropoda, which means 'jointed leg', all the animals have a skeleton on the outside of their body and have jointed legs. The phyla of the animal kingdom can be put into two groups called the invertebrates and the vertebrates. Invertebrates do not have an inside skeleton of cartilage or bone. Vertebrates do have an inside skeleton of cartilage or bone. The main phyla in the invertebrate group are the jellyfish, flatworms, annelid worms, arthropods, molluscs and echinoderms. There is only one phylum in the vertebrate group. It is known as the Chordata. The invertebrate and vertebrate groups are not actually part of the classification system but they are widely known and used to separate the animals in the Chordata from the animals in the other phyla.

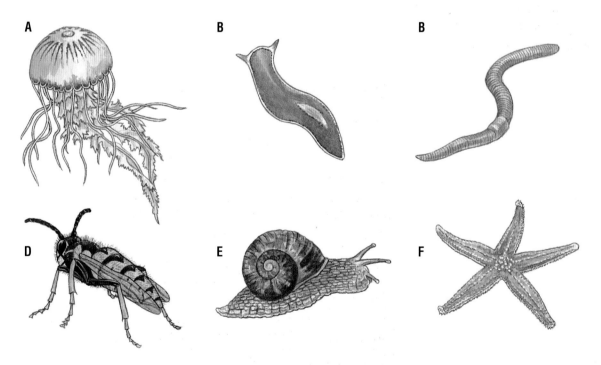

Figure 4.8 Examples of invertebrates. A) Jellyfish (compass jellyfish), B) flatworm (pond flatworm), C) annelid worm (earthworm), D) arthropod (wasp), E) mollusc (snail) and F) echinoderm (starfish).

6 How is an earthworm different from a wasp, and how is it similar to a wasp?

Each phylum is divided up into groups called classes. The phylum Chordata contains seven classes that form the vertebrate group. The classes are jawless fish, cartilaginous fish, bony fish, amphibians, reptiles, birds, and mammals.

Table 4.2 The features of five of the vertebrate classes.

Bony fish	Amphibians	Reptiles	Birds	Mammals
Scales, fins Eggs laid in water	Smooth skin Eggs laid in water	Scales. Soft-shelled eggs laid on land	Feathers Hard-shelled eggs	Hair Suckle young with milk

Each class is divided up into smaller groups called orders. The members of each order have so many features in common that they look alike and are easy to group. There are 19 orders of mammals. Four examples as shown in Table 4.3.

Table 4.3 Four orders of mammals.

Insectivores	Bats	Rodents	Whales
Small body Long snout	Small body Wings	Chisel-like teeth for gnawing	Flippers Tail with fins

An order is made up of smaller groups called families. The members of the different families look similar but there are differences. This can be seen by looking quite closely, as shown in the four families of whales.

7 How is a goldfish different from a frog, and how is it similar?

8 The insectivore in Table 4.3 is a shrew and the rodent is a mouse. How are they different and how are they similar to each other?

9 How is a beaked whale different from a sperm whale?

10 How is a beaked whale similar to a sperm whale?

Table 4.4 Four families.

Beaked whale	Sperm whale	Dolphin	White whale
Few teeth, small flippers	Large head, rounded back fin	Sickle-shaped flippers and back fin	No back fin, blunt snout

The members of a family have differences between them and are split up into smaller groups called genera (*singular:* genus). The differences between members of each genus are found by looking very closely. For example, if you look at dolphins A, B and C you will see that A seems to have more features in common with B than C. Because of this, dolphins A and B are placed in one genus and C is placed in a separate genus.

A Dusky dolphin

B White-sided dolphin

C Bottle-nosed dolphin

Figure 4.9 *Members of the dolphin family.*

11 How are dolphins A and B different from C?

12 How are dolphins A and B different from each other?

Because dolphins A and B have small differences between them, they are placed in separate groups called species. A species is a group of animals that have a very large number of similarities and the males and females breed together to produce offspring that can also breed. The males and females of different species do not normally breed together, but when they do they produce offspring that are usually sterile (cannot breed). For example, a male donkey and a female horse produce a sterile mule.

Setting up a system

In the 17th and 18th Centuries explorers were travelling the world in sailing ships and bringing back thousands of specimens of living things that had not been seen before by scientists. There was a great need for a system of grouping that everyone could understand.

Carl Linnaeus (1707–1778) travelled in Scandinavia, western Europe and England collecting and studying plants. He worked out a way of describing how one kind of living thing was different from another. He then began putting very similar living things into the same group. He gave each living thing two names. The first name was the name of its genus and the second name was its own specific name or species name. The names were made from Latin and Greek. These were two languages that scientists of every country learnt, so everyone could understand them. The words usually described the appearance of a living thing. For example, the genus and species name of the African clawed toad is *Xenopus laevis*. Xenopus is made from two Greek words – xenos meaning 'strange' and pous meaning 'foot'. The words refer to the toad's webbed hind foot which has three toes, each capped with a dark, sharp claw. The word 'laevis' is Latin for smooth and refers to the toad's smooth skin.

1 Why was there a need for grouping living things in the 17th and 18th Centuries?
2 What age was Carl Linnaeus when he died?
3 Why were Latin and Greek used to name living things?

For discussion

Why were the common names or local names not used in the scientific naming of plants and animals?

Other kingdoms of living things

As well as the animals, there are four other kingdoms of living things. They are the plants, Monera, Protoctista and fungi. You can find out about the plant kingdom in Chapter 8 and the other kingdoms in Chapter 7.

Keys

The way in which living things are divided up into groups can be used to identify them. The features of a phylum, class, order, family, genus or species can be set out as a spider key or organised as pairs of statements in a numbered key.

Spider key

On each 'leg' of the spider is a feature that is possessed by the living thing below it. An example is shown in Figure 4.10. A spider key is read by starting at the top in the centre and reading the features down the legs until the specimen is identified.

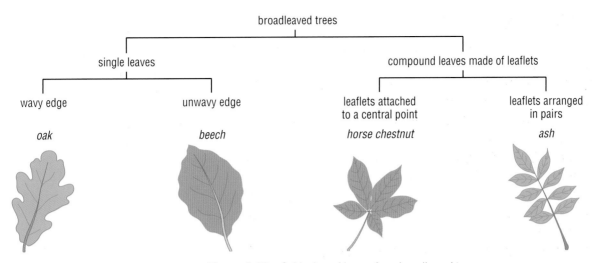

Figure 4.10   Spider key of leaves from broadleaved trees.

13 Make a spider key for the four animals in Figure 4.11. Look carefully at the animals. Choose a feature they have all got in common to start at the top of the key.

Figure 4.11   Molluscs and annelid worms.

Numbered key

You work through a numbered key by reading each pair of statements and matching the description of one of them to the features you see on the specimen you are trying to identify. At the end of each statement there is an instruction to move to another pair of statements or to the name of a living thing. Here is a simple numbered key. It can be used to identify molluscs that live in freshwater habitats such as rivers, lakes and ponds.

1 a) Single shell .... see 2
  b) Two shells .... see 6
2 a) Snail with a plate that closes
      the shell mouth .... *Bithynia*
  b) Snail without a plate that
      closes the shell mouth........................ see 3
3 a) Snail without a twisted shell .... Freshwater
      .... limpet
  b) Snail with a twisted shell.................... see 4
4 a) Shell in a coil .... Ramshorn
      .... snail
  b) Shell without a coil............................. see 5
5 a) Snail with triangular tentacles............ Pond snail
  b) Snail with long, thin tentacles............ Bladder snail
6 a) Animal has threads attaching
      it to a surface .... Zebra mussel
  b) Animal does not have threads
      attaching it to a surface...................... see 7
7 a) Shell larger than 25 mm .... Freshwater
      .... mussel
  b) Shell smaller than 25 mm .... Pea mussel

14 Identify the molluscs A–F in Figure 4.12 using the numbered key on page 75. In each case write down the number of each statement you used to make the identification.

15 Why should another feature in addition to size be added to the statements in part 7 of the key?

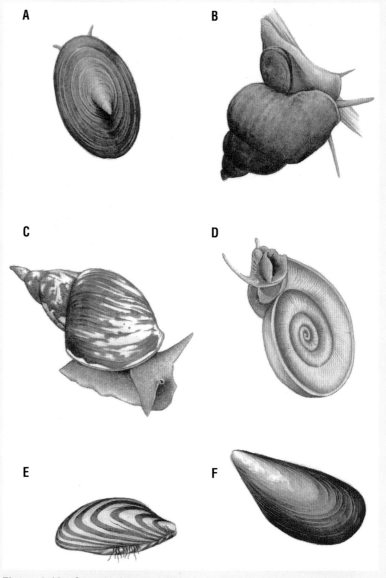

A

B

C

D

E

F

Figure 4.12 Some freshwater molluscs.

When using a numbered key write down the numbers of the statements you followed to identify your specimen. For example, specimen A is identified by following statements **1a**, **2b**, **3a**. It is a freshwater limpet.

16 Make up a numbered key to identify the arthropods in Figure 4.13. Begin by separating the butterfly, which has six legs, from the others.

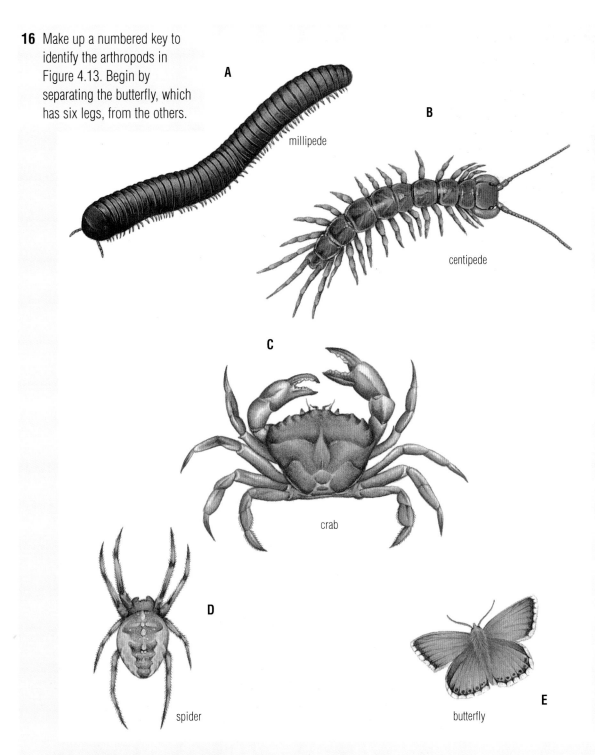

A millipede

B centipede

C crab

D spider

E butterfly

Figure 4.13 Arthropod specimens.

17 Look at the parts of the
insect's body in Figure 4.14a.
Then look at how the parts
vary among the six insect
specimens in b. Invent a key
to identify them.

18 Look at the numbered key for
freshwater molluscs and the
spider key for the leaves (both
on page 75) and answer the
following questions.

a) Which key identifies the
larger number of living
things?

b) If both keys featured the
same number of living
things, which key would
need the larger amount of
space?

c) What is an advantage of
the numbered key?

d) What is an advantage of a
spider key?

e) Which is the better one to
use in a poster? Explain.

f) Which is the better one to
use in a pocket book for
field work? Explain.

a)

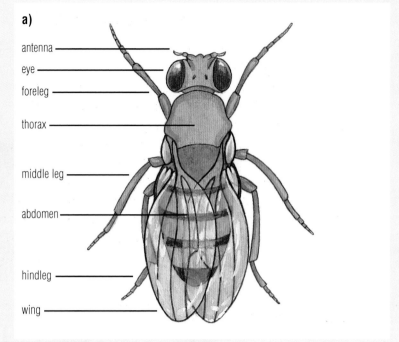

b)

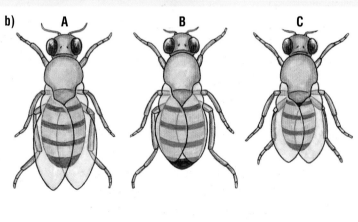

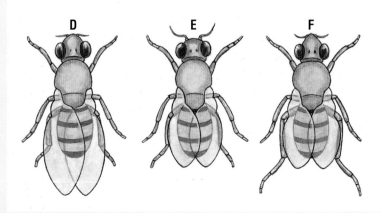

Figure 4.14 Variation in flies.

♦ SUMMARY ♦

♦ Detailed features of living things are seen by close observation and accurate recording (*see page 64*).
♦ There is variation between species (*see pages 65–66*).
♦ There are two kinds of variation within a species. These are continuous variation and discontinuous variation (*see page 67*).
♦ The environment can affect variation in a species (*see page 67*).
♦ There are seven characteristics of life. They are feeding, respiring, moving, growing, breeding, excreting and irritability (*see page 68*).
♦ Living things are classified by putting them into groups. The groups are kingdom, phylum, class, order, family, genus and species (*see pages 70–72*).
♦ Keys are used to identify living things (*see pages 74–75*).

End of chapter questions

1 What kind of living organisms are the following:
   a) does not have a backbone but has five arms,
   b) has a backbone and wings,
   c) does not have a backbone but has wings,
   d) has scales and lays eggs in water,
   e) has scales and lays eggs on land,
   f) has a backbone and hair?
2 Imagine that you have landed on a distant planet. When you climb out of your space craft you find some small eight-legged, six-eyed animals leaping about. You call them hoppys, and gather the following information about twenty of them.
   a) Arrange the hoppys into five size groups based on their mass.
   b) Display the numbers in the groups in a bar chart.
   c) Arrange the hoppys into groups based on colour.
   d) Display the numbers in the groups in a bar chart.
   e) Which feature shows continuous variation?
   f) Which feature shows discontinuous variation?
   g) Is there any relationship between the mass of the hoppys and their colours? Describe what you find.
   h) Speculate on a reason for your findings.

Table 4.5

Hoppy	Mass/g	Colour
1	200	green
2	349	green
3	210	green
4	615	blue
5	430	yellow
6	570	red
7	402	yellow
8	429	yellow
9	317	green
10	520	red
11	460	yellow
12	403	yellow
13	330	green
14	489	yellow
15	502	red
16	630	blue
17	410	yellow
18	380	green
19	550	red
20	445	yellow

5 Food and digestion

Figure 5.1 Food for sale at a school canteen.

1 Write a description of your daily eating pattern.
2 Compare your pattern with the two on this page. Which one does your pattern resemble?
3 From what you already know, try to explain which diet is healthier.

For discussion

How healthy is your eating pattern? What changes would make it healthier? Do other people agree?

Some people do not eat breakfast. They have some sweets or crisps on the way to school. At break they eat a chocolate bar or have a fizzy drink. At lunch time they always have chips with their meal. In the afternoon they have some more sweets and for their evening meal they avoid green vegetables. Through the evening they have snacks of sweets, crisps and fizzy drinks.

Other people eat a breakfast of cereals and milk, toast and fruit juice. They eat an apple at break and have a range of lunch time meals through the week which include different vegetables, pasta and rice. In the afternoon they may have an orange and eat an evening meal with green vegetables. They may have a milky drink at bed time.

Nutrients

A chemical that is needed by the body to keep it in good health is called a nutrient. The human body needs a large number of different nutrients to keep it healthy. They can be divided up into the following nutrient groups:

- carbohydrates
- proteins
- minerals
- fats
- vitamins

In addition to these nutrients the body also needs water. It accounts for 70% of the body's weight and provides support for the cells, it carries dissolved materials around the body and it helps in controlling body temperature.

Carbohydrates

Carbohydrates are made from the elements carbon, hydrogen and oxygen. The atoms of these elements are linked together to form molecules of sugar. There are different types of sugar molecule but the most commonly occurring is glucose. Glucose molecules link together in long chains to make larger molecules, such as starch. Glucose and starch are two of the most widely known carbohydrates but there are others, such as cellulose.

in starch, each of these links is a glucose molecule

Figure 5.2 Carbohydrate molecule.

Fats

Fats are made of large numbers of carbon and hydrogen atoms linked together into long chains together with a few oxygen atoms. There are two kinds of fats – the solid fats produced by animals, such as lard, and the liquid fat or oil produced by plants, such as sunflower oil.

Proteins

Proteins are made from atoms of carbon, hydrogen, oxygen and nitrogen. Some proteins also contain sulphur and phosphorus. The atoms of these elements join together to make molecules of amino acids. Amino acids link together into long chains to form protein molecules.

4 What elements are found in carbohydrates, fats and proteins?

5 Which two words are used to describe the structure of carbohydrate, fat and protein molecules?

6 A science teacher held up a necklace of beads to her class and said it was a model of a protein molecule. What did each bead represent?

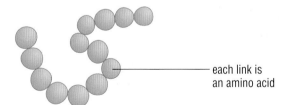

each link is an amino acid

Figure 5.3 Protein molecule.

Vitamins

Unlike carbohydrates, fats and proteins, which are needed by the body in large amounts, vitamins are only needed in small amounts. When the vitamins were first discovered they were named after letters of the alphabet. Later, when the chemical structure of their molecules had been worked out, they were given chemical names.

Minerals

The body needs 20 different minerals to keep healthy. Some minerals, such as calcium, are needed in large amounts but others, such as zinc, are needed in only tiny amounts and are known as trace elements.

INGREDIENTS			
MAIZE, SUGAR, MALT FLAVOURING, SALT, NIACIN, IRON, VITAMIN B$_6$, RIBOFLAVIN (B$_2$), THIAMIN (B$_1$), FOLIC ACID, VITAMIN B$_{12}$.			

NUTRITION INFORMATION		Typical value per 100g	Per 30g Serving with 125ml of Semi-Skimmed Milk
ENERGY	kJ	1550	700 *
	kcal	370	170
PROTEIN	g	8	7
CARBOHYDRATE	g	83	31
(of which sugars)	g	(8)	(9)
(starch)	g	(75)	(22)
FAT	g	0.7	2.5*
(of which saturates)	g	(0.2)	(1.5)
FIBRE	g	3	0.9
SODIUM	g	1.1	0.4
VITAMINS:		(%RDA)	(%RDA)
THIAMIN (B$_1$)	mg	1.2 (85)	0.4 (30)
RIBOFLAVIN (B$_2$)	mg	1.3 (85)	0.6 (40)
NIACIN	mg	15 (85)	4.6 (25)
VITAMIN B$_6$	mg	1.7 (85)	0.6 (30)
FOLIC ACID	μg	333 (165)	110 (55)
VITAMIN B$_{12}$	μg	0.85 (85)	0.75 (75)
IRON	mg	7.9 (55)	2.4 (17)

Figure 5.4 The nutrients in a food product are displayed on the side of the packet.

How the body uses nutrients

Carbohydrates

Carbohydrates contain a large amount of energy that can be released quickly inside the body. They are used as fuels to provide the energy for keeping the body alive. Cellulose, which makes up the walls of plant cells, is a carbohydrate. We cannot digest it but its presence in our food gives the food a solid property. This allows the muscles of the gut to push the food along, aiding digestion and preventing constipation. Cellulose in food is known as dietary fibre.

Fats

Fats are needed for the formation of cell membranes. They also contain even larger amounts of energy than carbohydrates. The body cannot release the energy in fats as quickly as the energy in carbohydrates, so fats are used to store energy. In mammals the fat forms a layer under the skin. This acts as a heat insulator and helps to keep the mammal warm in cool conditions. Many mammals increase their body fat in the autumn so that they can draw on the stored energy if little food can be found in the winter. Some plants store oil in their seeds.

Proteins

Proteins are needed for building the structures in cells and in the formation of tissues and organs. They are needed for the growth of the body, to repair damaged parts, such as cut skin, and to replace tissues that are constantly being worn away, such as the lining of the mouth.

Chemicals that take part in the reactions for digesting food and in speeding up reactions inside cells are called enzymes. These are also made from proteins.

Vitamins

Each vitamin has one or more uses in the body. Vitamin A is involved in allowing the eyes to see in dim light and in making a mucus lining to the respiratory, digestive and excretory systems which protects against infection from microorganisms.

There are several B vitamins of which vitamin B_1 (thiamin) is an example (see page 84).

A lack of vitamin C causes the deficiency disease called scurvy. As the disease develops bleeding occurs at the gums in the mouth, under the skin and into the joints. Death may occur due to massive bleeding in the body.

Vitamin D helps the body take up calcium from food to make strong bones and teeth. Children who have a lack of vitamin D in the diet develop the deficiency disease called rickets, in which the bones do not develop to their full strength and may therefore bend. This is seen particularly in the leg bones. Look at the X-ray in Figure 5.5.

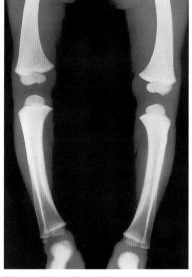

Figure 5.5 This child is suffering from rickets. It can be prevented by adding vitamin D to the diet.

Table 5.1 Vitamins and their uses.

Vitamin	Effect on body	Good sources
A	Increased resistance to disease Helps eyes to see in the dark	Milk, liver, cod-liver oil
B_1	Prevents digestive disorders Prevents the disease beriberi	Bread, milk
C	Prevents the disease scurvy in which gums bleed and the circulatory system is damaged	Blackcurrant, orange, lemon
D	Prevents the disease rickets in which bones become soft and leg bones of children may bend	Egg yolk, butter, cod-liver oil

Finding the cause of beriberi

Christiaan Eijkman (1858–1930) was a Dutch doctor who worked at a medical school in the East Indies in the late 19th Century. He investigated the disease called beriberi. In this disease the nerves fail to work properly and the action of the muscles becomes weak. All movements, especially walking, become difficult and as the disease progresses the heart may stop.

At this time other scientists had recently shown that microorganisms cause a number of diseases. It seemed reasonable to think that beriberi was also caused by a microorganism of some kind. Eijkman set up investigations to find it. He was not having any success. Then one day a flock of chickens that were kept at the medical school began to show the symptoms of beriberi.

Figure A   A flock of hens with beriberi.

Eijkman tested them for signs of the microorganisms that he believed were causing the disease. Again he had no success in linking the disease to the microorganisms but while he was studying the chickens they recovered from the disease. Eijkman began to search for a reason why they had developed the disease and also why they had recovered so quickly. He discovered that the chickens were usually fed on chicken feed (a specially prepared mixture of foods that kept them healthy). A cook who had been working at the medical school had stopped using the chicken feed and had fed the chickens on rice that had been prepared for the patients. This cook had left and a new cook had been employed who would not let the rice be fed to the chickens. The birds were once again fed on the chicken feed. When Eijkman fed the chickens on rice again they developed beriberi. When he fed them on chicken feed they recovered from the disease straight away.

The rice fed to the chickens and the patients was polished rice. This had had its outer skin removed and appeared white. Later work by scientists showed that the skin of the rice contained vitamin B_1 or thiamin. This vitamin is needed to keep the nerves healthy and prevent beriberi.

1 What are the symptoms of beriberi?

2 How serious is the disease?

3 Why did Eijkman begin by looking for microorganisms as a cause of beriberi?

4 In what way did chance play a part in the discovery of the cause of beriberi? Explain your answer.

5 Write down a plan of an investigation to check Eijkman's work on chickens and beriberi. How would you make sure it was fair and that the results were reliable?

6 How did Eijkman's work alter the way scientists thought diseases developed?

7 What is the danger in having a diet which mainly features polished rice? Explain your answer.

For discussion

Eijkman performed his experiments on animals. Question 5 asked you to plan an investigation to check his work. Your plan may have featured studying animals. A great deal of information that benefits humans has been gathered by studying animals in experiments. Are there any guidelines that you would want scientists to follow in experiments involving animals?

7 A meal contains carbohydrate, fat, protein, vitamin D, calcium and iron. What is the fate of each of these substances in the body?

8 Which carbohydrate cannot be digested by humans and how does it help the digestive system?

9 In a patient suffering from rickets why do the leg bones bend more than the arm bones?

10 A seal is a mammal. How can it survive in the cold polar seas when a human would die in a few minutes?

Figure 5.6 Seals on the ice.

Minerals

Each mineral may have more than one use. For example, calcium is needed to make strong bones and teeth. It is also used to make muscles work and for blood to clot. A lack of calcium in the diet can lead to weak bones and high blood pressure. The mineral iron is used to make the red blood pigment called haemoglobin.

Water

About 70% of the human body is water. The body can survive for only a few days without a drink of water.

Every chemical reaction in the body takes place in water. The blood is made mainly from water. It is the liquid that transports all the other blood components around the body.

Water is used to cool down the body by the evaporation of sweat from the skin.

The amounts of nutrients in food

The amounts of nutrients in foods have been worked out by experiment and calculation. The amounts are usually expressed for a sample of food weighing 100 g. Table 5.2 shows the nutrients in a small range of common foods.

Table 5.2 The nutrients in some common foods.

Food (100 g)	Protein (g)	Fat (g)	Carbohydrate (g)	Calcium (mg)	Iron (mg)	Vitamin C (mg)	Vitamin D (µg)
Potato	2.1	0	18.0	8	0.7	8–30	0
Carrot	0.7	0	5.4	48	0.6	6	0
Bread	9.6	3.1	46.7	28	3.0	0	0
Spaghetti	9.9	1.0	84.0	23	1.2	0	0
Rice	6.2	1.0	86.8	4	0.4	0	0
Lentil	23.8	0	53.2	39	7.6	0	0
Pea	5.8	0	10.6	15	1.9	25	0
Jam	0.5	0	69.2	18	1.2	10	0
Peanut	28.1	49.0	8.6	61	2.0	0	0
Lamb	15.9	30.2	0	7	1.3	0	0
Milk	3.3	3.8	4.8	120	0.1	1	0.05
Cheese 1	25.4	35.4	0	810	0.6	0	0.35
Cheese 2	15.3	4.0	4.5	80	0.4	0	0.02
Butter	0.5	81.0	0	15	0.2	0	1.25
Chicken	20.8	6.7	0	11	1.5	0	0
Egg	12.3	10.9	0	54	2.1	0	1.50
Fish 1	17.4	0.7	0	16	0.3	0	0
Fish 2	16.8	18.5	0	33	0.8	0	22.20
Apple	0.3	0	12.0	4	0.3	5	0
Banana	1.1	0	19.2	7	0.4	10	0
Orange	0.8	0	8.5	41	0.3	50	0

Notes for Tables 5.2 and 5.3

Vegetables are raw; the bread is wholemeal bread; cheese 1 is cheddar cheese; cheese 2 is cottage cheese; fish 1 is a white fish, such as cod; fish 2 is an oily fish, such as herring.

11 In Table 5.2, which foods contain the most
   a) protein,
   b) fat,
   c) carbohydrate,
   d) calcium,
   e) iron,
   f) vitamin C and
   g) vitamin D?
12 Which foods would a vegetarian not eat?
13 Which foods would a vegetarian have to eat more of and why?
14 Which food provides all the nutrients?
15 Why might you expect this food to contain so many nutrients?

Keeping a balance

In order to remain healthy the diet has to be balanced with the body's needs. A balanced diet is one in which all the nutrients are present in the correct amounts to keep the body healthy. You do not need to know the exact amounts of nutrients in each food to work out whether you have a healthy diet. A simple way is to look at a chart showing food divided into groups, with the main nutrients of each group displayed (see Table 5.4 page 88). You can then see if you eat at least one portion from each group each day and more portions of the food groups that lack fat. Remember that you also need to include fibre even though it is not digested. It is essential for the efficient working of the muscles in the alimentary canal. Fibre is found in cereals, vegetables and pulses, such as peas and beans.

Table 5.3 The energy value of some common foods.

Food (100 g)	Energy (kJ)
Potato	324
Carrot	98
Bread	1025
Spaghetti	1549
Rice	1531
Lentil	1256
Pea	273
Jam	1116
Peanut	2428
Lamb	1388
Milk	274
Cheese 1	1708
Cheese 2	480
Butter	3006
Chicken	602
Egg	612
Fish 1	321
Fish 2	970
Apple	197
Banana	326
Orange	150

16 Table 5.3 shows the amount of energy provided by 100 g of each of the foods shown in Table 5.2. Arrange the nine highest energy foods in order starting with the highest and ending with the lowest. Look at the nutrient content of these foods in Table 5.2.
   a) Do you think the energy is stored as fat or as carbohydrate in each of the nine highest energy foods?
   b) Arrange the foods into groups according to where you think the energy is stored.
   c) Do the food stores you have identified store the same amount of energy (see also page 82)? Explain your answer.
17 Why might people who are trying to lose weight eat cottage cheese instead of cheddar cheese?
18 Mackerel is an oily fish. Describe the nutrients you would expect it to contain.

19 Look again at the eating pattern you prepared for Question 1 on page 80. Analyse your diet into the food groups shown in Table 5.4. How well does your diet provide you with all the nutrients you need?

20 Table 5.5 shows how the energy requirements of an average male and female person change from the age of 2 to 25 years. Plot graphs of the information given in the table.

21 Describe what the graphs show.

22 Explain why there is a difference in energy needs between a 2-year-old child and an 8-year-old child.

23 Explain why there is a difference in energy needs between an 18-year-old male and an 18-year-old female.

24 Explain why there is a change in the energy needs as a person ages from 18 to 25.

25 What changes would you expect in the energy used by:
   a) a 25-year-old person who changed from a job delivering mail to working with a computer
   b) a 25-year-old person who gave up working with computers and took a job on a building site that involved carrying heavy loads
   c) a 25-year-old female during pregnancy?

26 What happens in the body if too much fat, carbohydrate or protein is eaten?

27 Why do people become thin if they do not eat enough high-energy food?

Table 5.4 The groups of foods and their nutrients.

Vegetables and fruit	Cereals	Pulses	Meat and eggs	Milk products
Carbohydrate	Carbohydrate	Carbohydrate	Protein	Protein
Vitamin A	Protein	Protein	Fat	Fat
Vitamin C	B vitamins	B vitamins	B vitamins	Vitamin A
Minerals	Minerals	Iron	Iron	B vitamins
Fibre	Fibre	Fibre		Vitamin C
				Calcium

Table 5.5 Average daily energy used by males and females.

Age (years)	Daily energy used (kJ)	
	Male	Female
2	5500	5500
5	7000	7000
8	8800	8800
11	10 000	9200
14	12 500	10 500
18	14 200	9600
25	12 100	8800

Malnutrition

If the diet provides too few nutrients or too many nutrients malnutrition occurs. Lack of a nutrient in a diet may produce a deficiency disease, such as scurvy or anaemia. Scurvy is a deficiency disease caused by a lack of vitamin C and anaemia is a deficiency disease caused by a lack of iron.

If more protein than is needed is eaten, it is broken down in the body. Part of it is converted to a carbohydrate called glycogen, which is stored in the liver, and part of it is converted to a chemical called urea, which is excreted in the urine.

Too much high-energy food such as carbohydrate and fat leads to the body becoming overweight. If the body is extremely overweight it is described as obese. If too little high-energy food is eaten the body becomes thin because it uses up energy stored as fat. Energy stored in protein in the muscles can also be used up.

The condition anorexia nervosa can lead to extreme weight loss and possibly death. It occurs mainly in teenage girls but is occurring increasingly in teenage boys and adult men and women. People suffering from anorexia nervosa eat very little and fear gaining weight. As soon as the condition is diagnosed, they need careful counselling to stand the best chance of making a full recovery.

A healthy diet

The body needs a range of nutrients to keep healthy (see pages 86–87) and everyone should eat a balanced diet to provide these nutrients. Regular eating of high-energy snacks, such as sweets, chocolate, crisps, ice-cream and chips, between meals unbalances the diet and can lead to the body becoming overweight, damage to the teeth (see pages 124–125) and ill-health. Overweight people have to make more effort than normal to move so they tend to take less exercise. In time this can affect the heart (see page 181).

High-energy snacks should be kept to a minimum so that the main meals of the day, which provide most of the essential nutrients, may be eaten. There are alternatives to high-energy snacks. These are fruits and raw vegetables, such as celery, tomatoes and carrots. In addition to being lower in energy they also provide more vitamins and minerals.

Digestion

Your food comes from the tissues of animals and plants. To enter the cells of your body the tissues have to be broken down. This releases the nutrients (carbohydrates, proteins, fats, minerals and vitamins). Some of them are in the form of long-chain molecules. They must be broken down into smaller molecules that dissolve in water and can pass through the wall of the gut. This process of making the food into a form that can be taken into the body is called digestion. It takes place in the digestive system, which is made up of the alimentary canal and organs such as the liver and pancreas.

The breakdown of food

There are two major processes in the breakdown of food. They are the physical and the chemical breakdown of food. Food is physically broken down from large objects into small objects in the mouth. Chemical breakdown begins in the mouth and continues along the alimentary canal.

The physical breakdown of food

The teeth play a major part in the physical breakdown of food. There are four kinds of teeth. The chisel-shaped incisor teeth are at the front of the mouth. These are for biting into soft foods like fruits. Next to the incisors are the canines. These are pointed, and in dogs and cats they form the fang teeth that are used for tearing into tougher food like meat. Humans do not eat much tough food so they use their canines as extra incisors. The premolars and the molars are similar in appearance. They have raised parts called cusps with grooves between them. They form a crushing and grinding surface at the back of the mouth. The action of the teeth breaks up the food into small pieces.

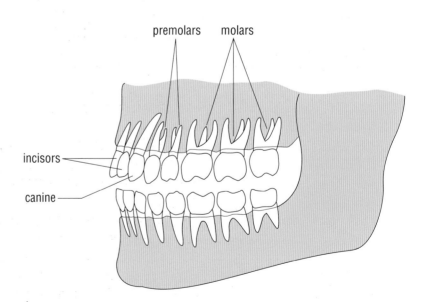

Figure 5.7   The four types of teeth in the mouth.

Early ideas about digestion

At one time there were two ideas about how food was digested. Some scientists believed that the stomach churned up the food to break it up physically and others believed that a chemical process took place.

Andreas Vesalius (1514–1564) was a Flemish doctor who investigated the structure of the human body by dissection. He had an artist make drawings of his work and these were published in a book for others to study.

Figure A A drawing of Vesalius's work.

1 How was Vesalius's work recorded?
2 How life-like were the recordings of Vesalius's work?
3 How did the idea that the stomach acted as a churning machine develop?
4 If Borelli's idea had been correct what would Réaumur have found?
5 What did Réaumur's investigation show?
6 Do you think Réaumur's investigation threatened the hawk's life? Explain your answer.

René Descartes (1596–1650) studied mathematics and astronomy. He believed that all actions were due to mechanical movements. When he saw the drawings of Vesalius's dissections he believed that the human body behaved just as a machine.

Giovanni Borelli (1608–1679) studied the parts of the body and Descartes's ideas. He showed how muscles pulled on bones to make them move and how the bones acted as levers. This work supported Descartes's ideas, and Borelli extended it to consider the stomach as a churning machine for breaking up food.

Franciscus Sylvius (1614–1672), a German doctor, believed that chemical processes took place in the body and that digestion was a chemical process that began in the mouth with the action of saliva. Some other scientists believed in his ideas.

Figure B A hawk eating a meal.

In 1752 René Réaumur (1683–1757), a French scientist, decided to test these two ideas by studying the digestion in a hawk. When a hawk feeds it swallows large pieces of its prey, digests the meat and regurgitates fur, feathers and bones that it cannot digest. Réaumur put some meat inside small metal cylinders and covered the ends with a metal gauze. He fed the cylinders to the hawk and waited for the hawk to regurgitate them. He found that some of the meat had dissolved but the cylinders and gauze showed no signs of being ground up as if by a machine. To follow up his experiment he fed a sponge to the hawk to collect some of the stomach juices. When the hawk regurgitated the sponge Réaumur squeezed out the stomach juices and poured them on to a sample of meat. Slowly the meat dissolved.

The chemical breakdown of food

Proteins, fats and carbohydrates are made from large molecules which are made from smaller molecules that are linked together. The large molecules do not dissolve in water and cannot pass through the lining of the digestive system into the body. The smaller molecules from which they are made, however, *do* dissolve in water and *do* pass through the wall of the digestive system. Almost all reactions in living things involve chemicals called enzymes. They are made by the body from proteins and they speed up chemical reactions. Digestive enzymes speed up the breakdown of the large molecules into smaller ones.

28 Which smaller molecules join together to form
   a) carbohydrates and
   b) protein (see also page 81)?
29 What do enzymes in the digestive system do?

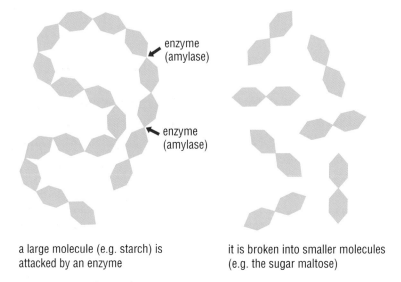

a large molecule (e.g. starch) is attacked by an enzyme

it is broken into smaller molecules (e.g. the sugar maltose)

Figure 5.8 The action of an enzyme on a large food molecule.

Along the alimentary canal

When your mouth waters

The 'water' that occurs in your mouth is called saliva. You can make up to $1\frac{1}{2}$ litres of saliva in 24 hours. Saliva is made by three pairs of salivary glands. The glands are made up of groups of cells that produce the saliva, and ducts (tubes) that deliver it to the mouth.

Saliva is 99% water but it also contains a slimy substance called mucin and an enzyme called amylase which begins the digestion of starch in the food. The mucin coats the food and makes it easier to swallow. Amylase begins the breakdown of starch molecules into sugar molecules.

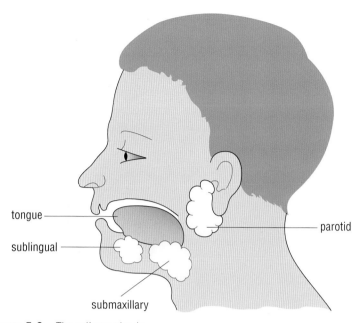

Figure 5.9   The salivary glands.

When you swallow

When you have chewed your food it is made into a pellet called the bolus. This is pushed to the back of your mouth by your tongue. Swallowing causes the bolus to slide down your gullet, which is the tube connecting the mouth to the stomach. This tube is also called the oesophagus. It has two layers of muscles in its walls.

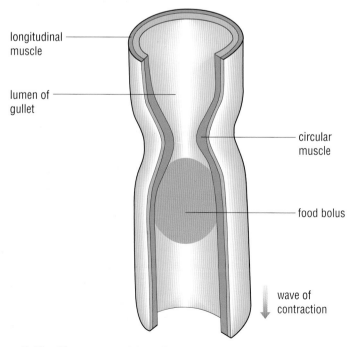

Figure 5.10   The structure of the gullet and the process of peristalsis.

In the outer layer the muscle cells are arranged so that they point along the length of the gullet. These form the longitudinal muscle layer. In the inner layer the cells are arranged so that they point around the wall of the gullet. These form the circular muscle layer.

Muscle cells can contract or get shorter. They cannot lengthen on their own, so another set of muscle cells must work to lengthen them. In the gullet, when the circular muscles contract, they squeeze on the food and push it along the tube. The longitudinal muscles then contract to stretch the circular muscles once again. The circular muscles do not all contract at the same time. Those at the top of the gullet contract first then a region lower down follows and so on until the food is pushed into the stomach. This wave of muscular contraction is called peristalsis. Peristaltic waves also occur in other parts of the alimentary canal to push the food along.

Stomach

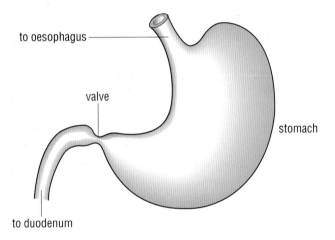

Figure 5.11 The stomach.

The stomach wall is lined with glands. These produce hydrochloric acid and a protein-digesting enzyme called pepsin. The hydrochloric acid kills many kinds of bacteria in the food and provides the acid conditions that pepsin needs to start breaking down protein in the food.

The food is churned up by the action of the muscles as they send peristaltic waves down the stomach walls at the rate of about three per minute. The food is prevented from leaving the stomach by a valve. When the food is broken down into a creamy liquid the valve opens which allows the liquid food to pass through into the next part of the digestive system.

30 How does saliva help in digesting food?

31 What is peristalsis?

32 What does hydrochloric acid do?

A hole in the stomach

In 1822 a group of fur trappers and hunters gathered at a trading post, Fort Mackinac, USA. One of the hunters accidentally fired his gun and shot a 19-year-old man called Alexis St Martin. It was fortunate that Doctor William Beaumont (1785–1853) was close by and could attend to the wounded man and save his life. St Martin had lost some flesh from over his stomach and part of the stomach wall. The wound did not completely heal. It formed a flap over the stomach which could be opened and the contents of the stomach examined.

St Martin agreed to help Beaumont to find out what happened inside the stomach during digestion. First Beaumont asked St Martin to eat nothing for a few hours then he looked inside the stomach and found that the stomach contained saliva, which St Martin had swallowed, and some mucus from the stomach wall.

In another experiment Beaumont put some bread crumbs into the stomach and saw digestive juice start to collect on the wall of the stomach.

Beaumont wanted to find out what happened to food in the stomach. So, he fastened pieces of cooked and raw meat, bread and cabbage onto silk strings and pushed them through the hole. An hour later he pulled the strings out and found that about half the cabbage and bread had broken up but the meat remained the same. Another hour later he found that the cooked meat had started to break down.

Next Beaumont wanted to find out what happened to the food after St Martin had eaten it. He gave St Martin a meal of fish, potatoes, bread and parsnips. After half an hour Beaumont examined the stomach contents and found that he could still identify pieces of fish and potato. After another half hour pieces of potato could still be seen but most of the fish had broken up. One and a half hours after the meal all the pieces of the food had broken up. Two hours after the meal the stomach was empty.

1 In the first experiment Beaumont was interested to find out if the stomach contained digestive juices all the time, even when no food was present.
  a) What conclusion could he draw from his observation?
  b) What prediction could he make from this observation?
2 What do you think Beaumont concluded from his second experiment?
3 The juices contain important chemicals made by the body. What are these called?
4 What did you think Beaumont concluded from his experiments with food on strings?
5 What do you think Beaumont concluded from his experiment on St Martin's meal?
6 Beaumont also investigated the action of the stomach juice outside the stomach. Why would he have kept the juice at body temperature?
7 If you were Alexis St Martin would you have allowed Dr Beaumont to carry out his investigation? Explain your reasons.

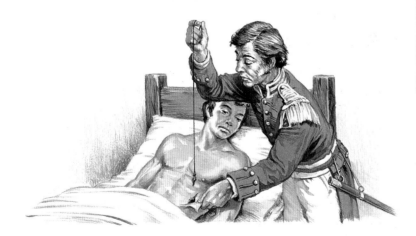

Figure A Doctor Beaumont placing a piece of food into Alexis St Martin's stomach.

Duodenum, liver and pancreas

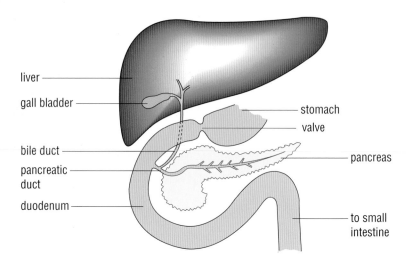

Figure 5.12   The duodenum, liver and pancreas.

The duodenum is a tube that connects the stomach to the small intestine. Two other tubes are connected to it. One tube carries a green liquid called bile from the gall bladder to mix with the food. Bile is made in the liver and contains chemicals that help break down fat into small droplets so that fat-digesting enzymes can work more easily. The second tube comes from an organ called the pancreas. This is a gland that produces a juice containing enzymes that digest proteins, fats and carbohydrates. The mixture of liquids from the stomach, liver and pancreas pass on into the small intestine.

Small intestine

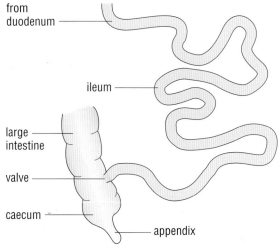

Figure 5.13   The small intestine.

33 Where is bile made and what does it do?

34 What are proteins, fats and carbohydrates broken down into?

35 Where are the digested foods absorbed?

The cells lining the wall of the small intestine make enzymes that complete the digestion of carbohydrates and proteins. Proteins are broken down into amino acids, carbohydrates are broken down into sugars, and fats are broken down into fatty acids and glycerol. All these small molecules are soluble and can pass through the wall of the small intestine. They are carried by the blood to all cells of the body.

Fate of undigested food

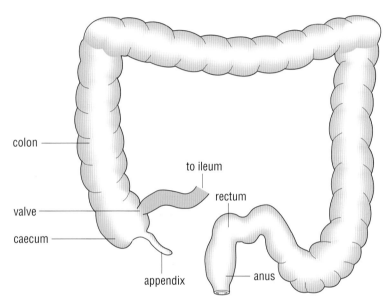

Figure 5.14 The large intestine and rectum.

36 What happens to undigested food in the large intestine?

37 What happens in egestion?

Indigestible parts of the food, such as cellulose, pass on through the small intestine to the large intestine and colon. Here water and some dissolved vitamins are absorbed and taken into the body. The remaining semi-solid substances form the faeces which are stored in the rectum. The faeces are removed from the body through the anus perhaps once or twice a day in a process called egestion.

38 What kind of enzyme is produced in
   a) the mouth and
   b) the stomach?
39 What kind of enzyme does bile help?
40 Where does bile come from?
41 Which organ of the digestive system produces all three kinds of enzyme?
42 Why do small droplets of fat get broken down by enzymes more quickly than large droplets?

Enzymes

An enzyme that digests carbohydrate is called a carbohydrase. An enzyme that digests protein is called a protease. An enzyme that digests fat is called a lipase.

Table 5.6 Enzymes.

Region of production	Kind of enzyme	Notes
Salivary glands in mouth	Carbohydrase	Enzyme is called salivary amylase
Gastric glands in stomach	Protease	Enzyme is called pepsin Hydrochloric acid is also made to help the enzyme work
Pancreas	Protease, carbohydrase, lipase	Enzymes enter the duodenum and mix with food and bile

◆ SUMMARY ◆

◆ A chemical that is needed by the body to keep it healthy is called a nutrient (*see page 80*).
◆ The groups of nutrients are carbohydrates, fats, proteins, vitamins and minerals (*see page 80*).
◆ Each nutrient has a specific use in the body (*see pages 82–83, 85*).
◆ Different foods have different amounts of nutrients (*see page 86*).
◆ A balanced diet needs to be eaten for good health (*see page 87*).
◆ Water and fibre are essential components of the diet (*see pages 85, 87*).
◆ The purpose of digestion is to break down the food into substances that can be absorbed and used by the body (*see page 89*).
◆ There are four kinds of teeth. They are the incisors, canines, premolars and molars. They have special shapes for specific tasks (*see page 90*).
◆ Enzymes break down the large molecules in food into smaller molecules so that they can be absorbed by the body (*see page 92*).
◆ The food is moved along the gut by a wave of muscular contraction called peristalsis (*see page 93*).
◆ The food is digested by enzymes that are made in the salivary glands, the stomach wall, the pancreas and the wall of the small intestine (*see pages 92, 94–95*).
◆ The liver produces bile which helps in the digestion of fat (*see page 96*).
◆ Digested food is absorbed in the small intestine (*see pages 96–97*).
◆ The undigested food has water removed from it in the large intestine and is then stored in the rectum before being released through the anus (*see page 97*).

End of chapter questions

1 What is a healthy diet?
2 Collect a copy of Figure 5.15 from your teacher and use the other diagrams in this chapter to label all the parts of the digestive system.
3 Describe the digestion of a chicken sandwich.

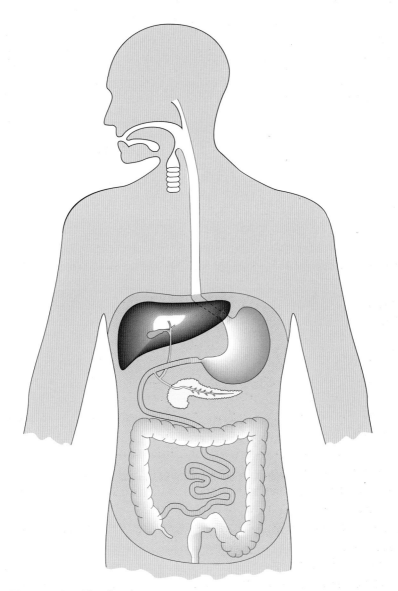

Figure 5.15   The digestive system.

(continued)

4 Do animals prefer certain foods? A pair of zebra finches were tested with a seed mixture bought from a pet shop to see if they preferred to eat certain seeds in the mixture. A sample of the mixture was left in a dish in the birds' cage for 6 hours. At the end of that time the sample was removed and the seeds were separated into their different types. A sample of the original mixture, called the bulk, that was similar in size to the dish sample was also sorted into the different seed types.

Table 5.7

	Type A Millet	Type B Round brown seeds	Type C Elongate grey seeds	Type D Small round seeds	Type E Black seeds
Dish (total sample = 218)	25	39	27	124	3
Bulk (total sample = 288)	120	27	41	97	3

Table 5.7 shows the composition of the two samples. Table 5.8 shows the percentage of each type of seed in the two samples.

Table 5.8

	Type A	Type B	Type C	Type D	Type E
Dish (D)	12	14	13	59	1.5
Bulk (B)	42	9	14	34	1
Difference (D − B)	−30	+5	−1	+25	−0.5

a) Why was the dish sample left for 6 hours in the bird cage?

b) How was the test made fair?

c) Why could the figures for the seeds in the two samples in Table 5.7 not be compared directly?

d) How is the percentage of the seed type worked out?

e) Check the percentage of millet in both samples. How have the figures been processed?

f) If the dish sample had roughly the same composition as the bulk sample when it was first put in the birds' cage,
   i) which seeds have the birds eaten, and
   ii) which have they most strongly avoided?

g) How could you find out more about the birds' food preferences?

6 Respiration

Respiration

All life processes require energy. The energy is stored in food molecules and is released in respiration. In the human body a sugar called glucose is the main source of energy. Most of it is formed by the digestion of starch. It dissolves in the blood and is transported to the cells.

Detecting the release of energy

When energy changes from stored energy in food to movement energy for example, some energy is released as heat. This is why you become hot when you take part in sport.

The production of heat can be used to indicate that something is alive. Seeds store energy and materials for plant growth. When seeds take in water the stored energy is used to power the growth of the plant inside, and germination occurs.

The production of heat by germinating seeds can be demonstrated in the following way. Two flasks are set up as shown in Figure 6.1.

1 Would you expect a rise in temperature if you put dead mouldy peas in a flask? Explain your answer.
2 In the burning process, a fuel takes part in a chemical reaction with oxygen and produces the same products as in respiration. How is burning different from respiration?
3 What would happen if burning occurred in the body instead of respiration?

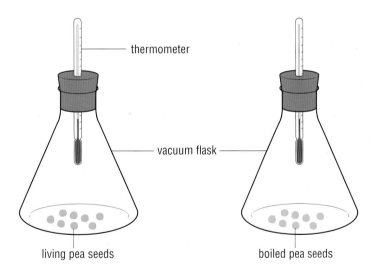

Figure 6.1 Investigating heat produced by germinating seeds.

After 24 hours the temperature inside each flask is measured. The flask containing the living pea seeds will be found to be at a higher temperature than the other flask. This indicates that the living pea seeds are respiring.

Aerobic respiration

Glucose takes part in a chemical reaction with oxygen inside the cell. During this reaction glucose is broken down to carbon dioxide and water, and energy is released. This process is called aerobic respiration. The reaction can be written as a word equation:

glucose + oxygen → carbon dioxide + water + energy

The energy is released slowly in a series of stages during respiration.

Respiratory system

In Chapter 5 we saw how food is broken down in the digestive system in readiness for transport to the cells. In this section we look at how the respiratory system provides a means of exchanging the respiratory gases – oxygen and carbon dioxide.

The function of the respiratory system is to provide a means of exchanging oxygen and carbon dioxide that meets the needs of the body, whether it is active or at rest. In humans the system is located in the head, neck and chest. It can be divided into three parts – the air passages and tubes, the pump that moves the air in and out of the system, and the respiratory surface. The terms respiration and breathing are often confused but they do have different meanings. Breathing just describes the movement of air in and out of the lungs. Respiration covers the whole process by which oxygen is taken into the body, transported to the cells and used in a reaction with glucose to release energy, with the production of water and carbon dioxide as waste products.

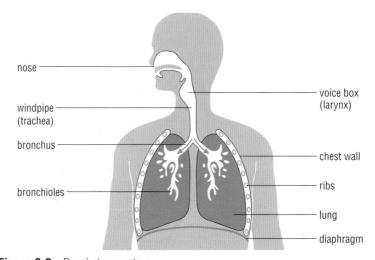

Figure 6.2 Respiratory system.

Air passages and tubes

Nose

Air normally enters the air passages through the nose. Hairs in the nose trap some of the dust particles that are carried on the air currents. The lining of the nose produces a watery liquid called mucus. This makes the air moist as it passes inwards and also traps bacteria that are carried on the air currents. Blood vessels beneath the nasal lining release heat that warms the air before it passes into the lungs.

Windpipe

The windpipe or trachea is about 10 cm long and 1.5 cm wide. It is made from rings of cartilage which is a fairly rigid substance. Each ring is in the shape of a 'C'. The inner lining of the windpipe has two types of cells. They are mucus-secreting cells and ciliated epithelial cells. Dust particles and bacteria are trapped in the mucus. The cilia beat backwards and forwards to move the mucus to the top of the windpipe where it enters the back of the mouth and is swallowed.

Bronchi and bronchioles

4 What structures hold the air passages open in the windpipe and bronchi?

5 Why is it more difficult to breathe during an asthmatic attack?

The windpipe divides into two smaller tubes called bronchi. (This is the name for more than one tube. A single tube is called a bronchus.) The two bronchi are also made of hoops of cartilage and have the same lining as the windpipe.

The bronchi divide up into many smaller tubes called bronchioles. These have a diameter of about 1 mm. The bronchioles divide many times. They have walls made of muscle but do not have hoops of cartilage. The wall muscles can make the bronchiole diameter narrower or wider.

Some people suffer from asthma. They may be allergic to certain proteins in food or to the proteins in dust that come from fur and feathers. The presence of these proteins in the air affects the muscles in the bronchioles, and the air passages in the bronchioles become narrower. This makes breathing very difficult. A person suffering an asthmatic attack can use an inhaler that releases chemicals to make the muscles relax to widen the bronchioles.

Air pump

The two parts of the air pump are the chest wall and the diaphragm. They surround the cavity in the chest. Most of the space inside the chest is taken up by the lungs. The outer surfaces of the lungs always lie close to the inside wall of the chest. The small space between the lungs and the chest wall is called the pleural cavity. The cavity contains a film of liquid that acts like a lubricating oil, helping the lung and chest wall surfaces to slide over each other during breathing.

6 Why is there a film of liquid in the pleural cavity?

Chest wall

This is made by the ribs and their muscles. Each rib is attached to the backbone by a joint that allows only a small amount of movement. The muscles between the ribs are called the internal and external intercostal muscles. The action of these muscles moves the ribs.

Diaphragm

This is a large sheet of muscle attached to the edges of the 10th pair of ribs and the backbone. It separates the chest cavity, which contains the lungs and heart, from the lower body cavity, which contains the stomach, intestines, liver, kidneys and female reproductive organs.

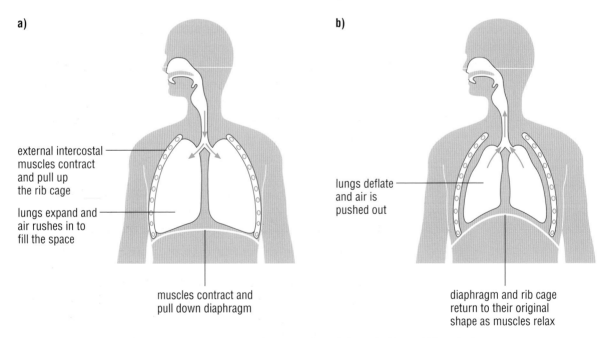

a)

external intercostal muscles contract and pull up the rib cage

lungs expand and air rushes in to fill the space

muscles contract and pull down diaphragm

b)

lungs deflate and air is pushed out

diaphragm and rib cage return to their original shape as muscles relax

Figure 6.3   Illustration of **a)** inspiration and **b)** expiration.

Breathing movements

There are two breathing movements – inspiration and expiration.

Inspiration

During inspiration the external intercostal muscles contract and the ribs move upwards and outwards. The muscles of the diaphragm contract, pulling it down into a flatter position. These actions increase the volume of the chest and reduce the pressure of air inside it. Therefore, air rushes in through the trachea and bronchi from outside the body. It is pushed in by the pressure of the air outside the body.

Expiration

The external intercostal muscles relax and the ribs fall back into their original position. Gravity is the main force that lowers the ribs and moves them inwards but the weak internal intercostal muscles may also help when they contract. The muscle fibres in the diaphragm relax and it rises to its dome-shaped position again. The organs below the diaphragm, which were pushed down when the diaphragm muscles contracted, now push upwards on the diaphragm. As the volume of the chest decreases, the pressure of the air inside it increases and air is pushed to the outside through the air passages.

Depth of breathing

The amount of air breathed in and out at rest is called the tidal volume and is about 500 cm^3 in humans. The maximum amount of air that can be breathed in and out is called the vital capacity. In human adults the vital capacity may reach 4000 cm^3.

Respiratory surface

At the end of each bronchiole is a very short tube called the alveolar duct. Bubble-like structures called alveoli open into this duct. Each alveolus has a moist lining, a thin wall and is supplied with tiny blood vessels called capillaries.

Oxygen from the inhaled air dissolves in the moist alveolar lining and moves by diffusion through the walls of the alveolus and the capillary next to it. The oxygen diffuses into the blood and enters the red blood cells (see page 108), which contain a dark red substance called haemoglobin. The oxygen then combines with the haemoglobin to make oxyhaemoglobin, which is bright red. Blood that has received oxygen from the air in the lungs is known as oxygenated blood.

7 How does the action of the external intercostal and diaphragm muscles draw air up your nose?

8 How do the values of the tidal volume and vital capacity compare?

9 A resting person gets up and starts running. Describe two ways in which the person's breathing pattern changes.

10 Why should a person's breathing pattern change between resting and running?

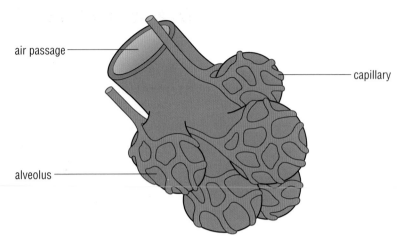

Figure 6.4 An alveolus.

Carbon dioxide is dissolved in the watery part of the blood called the plasma. It moves by diffusion through the capillary and alveolar walls and changes into a gas as it leaves the moist lining of the alveolus.

Blood moves through the capillaries very quickly, so a large amount of oxygen and carbon dioxide can be exchanged in a short time.

The spongy structure of the lungs is produced by the 300 million alveoli which make a very large surface area through which the gases can be exchanged. It is like having the surface area of a tennis court wrapped up inside two footballs! If this surface area is reduced then health suffers (see Smoking and health, page 174).

11 How would thick-walled alveoli affect the exchange of the respiratory gases?

12 Compare the way the blood carries oxygen and carbon dioxide.

13 How do you think you would be affected if the surface area of your lungs was reduced?

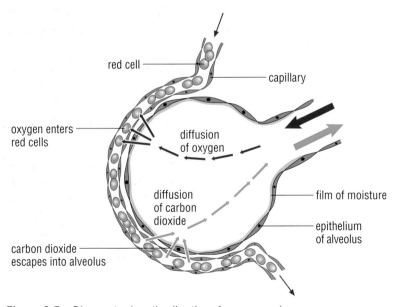

Figure 6.5 Diagram to show the direction of gaseous exchange.

What is in the blood?

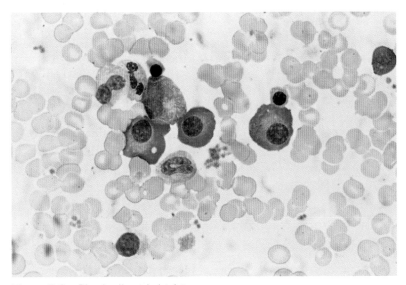

Figure 6.6 Blood cells and platelets.

14 Could you live without haemoglobin? Explain your answer.

15 Compare the tasks of red and white blood cells.

About 45% of a drop of blood is made from cells. There are two kinds, red cells and white cells.

Red cells contain haemoglobin which transports oxygen from the lungs to the other body cells. Haemoglobin allows the blood to carry 100 times more oxygen than the same amount of water. There are 500 red cells for every white cell.

White cells fight disease. They attack bacteria and produce chemicals to stop virus infections (see page 125). White cells also gather at the site of a wound where the skin has been cut. They eat bacteria that try to enter. The white cells die in this process and their bodies collect to form pus in the wound.

The blood also carries platelets which are fragments of cells. These collect in the capillaries at the site of a wound and act to block the flow of blood. Platelets help the blood to form clots at the site of a wound. These clots stop blood leaking out of the wound.

About 55% of blood is a watery liquid called plasma. This contains digested foods, hormones, such as adrenaline (see page 2), a waste product from the liver called urea and the carbon dioxide produced by all the body cells.

Moving oxygen to the cells

Once oxygen has entered the red blood cells it begins its journey inside the body. It starts by moving through the capillaries in the lungs. Capillaries are just one of three kinds of blood vessels found in the body (the other two are arteries and veins). They make a network of fine tubes in organs which provide a very large surface area between the blood and the tissues in the organs. This surface area allows a large amount of substances, such as oxygen and glucose, to pass between the blood and the tissues in a short amount of time. When the blood moves away from the lungs, it travels along a larger blood vessel called the pulmonary vein. The oxygenated blood is transported to the heart in the pulmonary vein, and is then circulated around the body.

Blood vessels

Arteries

Blood vessels that take blood away from the heart are called arteries. The high pressure of blood pushes strongly on the thick, elastic artery walls. They stretch and shrink as the blood moves by. This movement of the artery wall makes a pulse. When an artery passes close to the skin the pulse can be felt and therefore used to count how fast the heart is beating.

Veins

Blood vessels that bring blood towards the heart are called veins. The blood is not under such high pressure and so does not push as strongly on the vein walls. Veins have thinner walls than arteries and contain valves that stop the blood flowing backwards.

16 Why do veins have valves?

17 Why do you think arteries function better with thick walls?

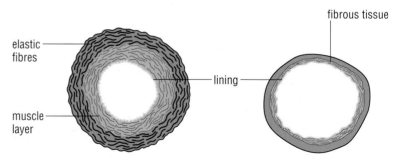

Figure 6.7   An artery and a vein.

Capillaries

When an artery reaches an organ it splits into smaller and smaller vessels. The smallest blood vessels are called capillaries. A capillary wall is only one cell thick. They are spread throughout the organ so that all cells have blood passing close to them. Where the blood leaves an organ, the capillaries join together to form larger and larger vessels until eventually they form veins.

The heart

Movement of the blood is produced by the pumping action of the heart. The heart is divided down the middle into two halves. The right side receives deoxygenated blood from the body and pumps it to the lungs. At the same time the left side receives oxygenated blood from the lungs and pumps it to the body.

The two pumps in the heart and a simplified arrangement of the blood vessels are shown in Figure 6.8.

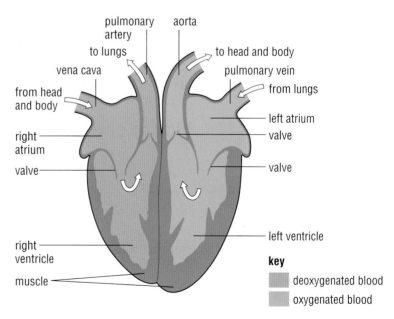

Figure 6.8 A simplified section through the heart.

18 Where does the pushing force come from to push the blood out of the heart?

19 What is the purpose of the heart valves?

20 Why do you think the walls of the left ventricle are thicker than the walls of the right ventricle?

Each side of the heart has two chambers. Blood flows from the veins into the upper chambers called the atria (*singular:* atrium). It passes from the atria into the lower chambers, the ventricles. The muscular walls of the ventricles relax as they fill up. When these muscular walls contract the blood is pumped into the arteries.

Valves between the atria and the ventricles stop the blood going backwards into the atria. Valves between the arteries (aorta and pulmonary artery) and the ventricles stop the blood from flowing backwards after it has been pumped out of the heart.

When the blood from the lungs leaves the heart it enters the aorta. This branches to other arteries which connect to different organs of the body. When the blood reaches an organ, it travels into another network of capillaries. It is here that the oxygen leaves the blood, passes through the wall of the capillary and enters the cells where respiration takes place.

Moving glucose to the cells

Glucose contains the store of energy that is released during respiration. It reaches the cells in the following way. Glucose passes through the wall of the small intestine and into the capillaries. It does not enter the red blood cells but stays in the yellow watery part of the blood called the plasma. The glucose travels in the plasma along veins which bring it to the heart. It enters the right side of the heart then passes to the lungs where the blood picks up oxygen. It then passes through the left side of the heart and into the aorta. From here it travels along arteries which take the glucose to the body organs.

Moving carbon dioxide to the lungs

Carbon dioxide is a product of aerobic respiration. It leaves the cells where it is produced and passes through the walls of the capillaries. It does not enter the red blood cells but stays in the plasma. It travels along veins which take it back to the right side of the heart. From here it enters the pulmonary artery and travels to the lungs. It escapes through the capillary walls into the air in the alveoli.

21 Make a simple drawing of a heart. Above the heart make a simple drawing of the lungs. Below the heart make a simple drawing of a muscle. Now using the information about how the respiratory gases travel, draw in blood vessels connecting the three organs.

Studies on the circulatory system

Erasistratus (about 304–250 BC) was a Greek doctor who studied the circulatory system. He suggested that veins and arteries carried different substances. He thought that veins carried blood and arteries carried 'animal spirit'.

Galen (about AD130–200) was also a Greek doctor. He used the pulse of a patient to help him to assess their sickness. He realised that the blood from one side of the heart got to the other side but he did not know how it happened. He thought there were tiny holes in the wall between the two sides of the heart. Galen also thought that the blood went backwards and forwards along the blood vessels. His ideas were held in high regard for over 1400 years.

Michael Servetus (1511–1553) was a Spanish doctor who traced the path of blood to and from the heart along the vein and artery that go to and from the lungs. He did not think that the blood went into the heart's muscular walls.

Fabricius ab Aquapendente (1537–1619) was a professor of surgery who discovered that the veins had valves in them. He taught the Englishman William Harvey (1578–1657) who became a doctor and went on to do further studies of the circulatory system. Fabricius's discovery of the valves gave Harvey a clue as to how the blood might flow. He followed up Fabricius's discovery by blocking an artery by tying a cord around it. He found that the side towards the heart swelled up because of the collecting blood. Next, he tied a cord around a vein. He found that the vein swelled on the side away from the heart.

Harvey also calculated the amount of blood that the heart pumped out in an hour. It was three times the weight of a man, yet the body did not increase in size. One explanation was that the heart made this amount of blood in an hour and another organ in the body destroyed it so the body did not increase in size. Harvey thought it impossible for the blood to be made and destroyed so quickly and so suggested that the blood must move around the body in only one direction. He published his ideas in a book in 1628 and was ridiculed by other doctors for challenging the ideas of Galen. Eventually the idea of the blood circulating round the body was accepted but Harvey could not explain how the blood got from the arteries to the veins.

Marcello Malpighi (1628–1694) was an Italian scientist who studied the wing of a bat under a microscope. He found that there was a connection between the arteries and veins in the wing. These were tiny vessels that could not be seen with the eye. These vessels were called capillaries and the blood could be seen flowing through them.

Figure A   William Harvey at work.

1 Who first described arteries and veins?
2 Who first began to doubt Galen's ideas?
3 How did Fabricius's discovery help Harvey?
4 If the blood flowed as Galen suggested what would Harvey have found when he tied off the artery and vein?
5 How did Harvey interpret his observations?
6 Why was Harvey's idea ridiculed?
7 How did Malpighi's work support Harvey's ideas?

For discussion

How could you adapt the apparatus shown in Figure 6.9 to find out if

i) other animals produce carbon dioxide, and if

ii) plants produce carbon dioxide?

Testing for carbon dioxide

In exhaled breath

Exhaled air can be tested for carbon dioxide by passing it through limewater. If carbon dioxide is present it reacts with the calcium hydroxide dissolved in the water to produce insoluble calcium carbonate. This makes the water turn white or milky.

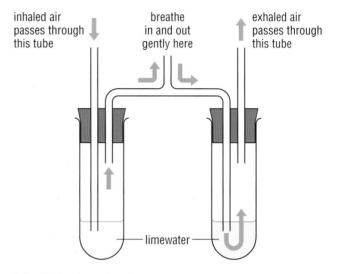

inhaled air passes through this tube

breathe in and out gently here

exhaled air passes through this tube

limewater

Figure 6.9 Testing inhaled and exhaled air for carbon dioxide.

In air around seeds

Carbon dioxide production can be used as an indication of respiration and a sign of life. Hydrogen carbonate indicator is a liquid that changes colour in the presence of carbon dioxide. It changes from an orange-red colour to yellow. The production of carbon dioxide by germinating pea seeds can be shown by setting up the apparatus shown in Figure 6.10.

For discussion

The apparatus shown in Figure 6.10 could be used to show that maggots release carbon dioxide. Should animals be used in experiments to show signs of life such as respiration?

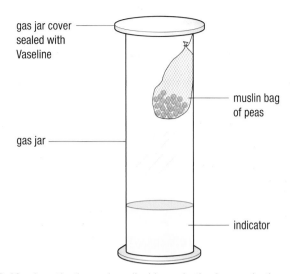

gas jar cover sealed with Vaseline

muslin bag of peas

gas jar

indicator

Figure 6.10   Investigating carbon dioxide production by germinating pea seeds.

Respiring without oxygen

22 If you are in good health, put one arm straight up in the air. Clench and unclench your fist as fast as you can for about 20 seconds. What happens and why? Put your arm down. What happens now and why?

Muscle cells sometimes need oxygen faster than it can be supplied. This might happen when you sprint for the finishing tape in a race. The muscle cells still respire, but they use another method. They respire anaerobically – without oxygen. In this process the glucose is broken down to lactic acid. As the cells respire anaerobically, the amount of lactic acid in the muscle increases and makes it ache. The lactic acid is broken down in the liver after being carried there in the blood.

◆ SUMMARY ◆

◆ Energy is released from food during respiration (*see page 101*).
◆ Oxygen and carbon dioxide are exchanged in the respiratory system (*see page 102*).
◆ Breathing is the process of air exchange between the air and the lungs (*see page 102*).
◆ Breathing movements, inspiration and expiration, are caused by the movement of the chest wall and the diaphragm (*see page 105*).
◆ Oxygen travels through the blood in the red blood cells (*see page 108*).
◆ The heart contains two pumps for moving the blood (*see page 109*).
◆ Glucose travels through the blood in the plasma (*see page 110*).
◆ Carbon dioxide travels through the blood in the plasma (*see page 110*).
◆ Carbon dioxide can be detected by indicators (*see page 112*).

End of chapter question

1 Four people took their pulses (measured in beats per minute, on a portable heart monitor) at rest, straight after exercise, one minute after exercise, two minutes after exercise and three minutes after exercise. Here are their results.

Andrew 71, 110, 90, 79, 71
Brenda 74, 115, 89, 77, 73
Clare 73, 125, 115, 108, 91
David 53, 80, 71, 46, 64

a) Make a table of the results.
b) Plot a graph of the results.
c) What trend can you see in the results?
d) When did Andrew's and Brenda's hearts beat at the same rate?
e) Andrew claims to be fitter than Clare. Do you think the results support his claim? Explain your answer.
f) Which result does not follow the trend? Explain why this may be so.

For discussion

Giorgio Baglivi (1668–1707) was an Italian doctor who believed that the body is just a machine. He matched scissors to teeth, bones to levers and lungs to bellows.

a) If he had been alive today, what might he have matched the brain to?

b) Was he right to think of the body as just a machine?

7 *Microbes*

Microorganisms

A microbe is a very tiny living organism and a microscope is needed to see it. The word 'microbe' does not refer to a group of living things in the same way that 'mammal' and 'bird' does. Microbes belong to three kingdoms. They are the fungi, Monera and Protoctista.

Fungi

Unlike plants, the cell wall of a fungus contains a substance called chitin. Many kinds of fungi can form very thin threads called hyphae. The hyphae feed on the dead bodies of plants and animals. As the hyphae grow they form a network called a mycelium. When the mycelium reaches a certain size it forms structures, called sporangia, that release spores. A spore is a reproductive cell with a thick wall around it. The wall provides protection from changes in temperature; it prevents the cell from losing water in dry conditions and gaining water in wet conditions. In the white mould that grows on bread, the sporangia are black spheres. Each one grows on a thread from the mycelium.

In some fungi the threads join together to make a larger structure called a fruiting body. This may be divided into a stalk and cap as seen in mushrooms and toadstools. There are spore-producing structures called gills in the cap. The mycelium, which produces fruiting bodies, feeds in the soil or in rotting wood and is not usually visible.

A few fungi are parasites (feed on living things). Athlete's foot (see page 131) is caused by a fungus that feeds on the skin between toes.

Yeasts are fungi that do not produce hyphae. 'Wild' yeasts feed on the sugar that forms on the surfaces of fruits and in the nectar of flowers. From these 'wild' yeasts, special types have been developed for use in baking and making alcoholic drinks.

1 Why are fungi not placed in the plant group?

Figure 7.1 Fairy-ring mushrooms. The fruiting bodies have been made at the edges of a disc-shaped mycelium in the soil.

Useful yeast

Yeast respires anaerobically (see page 113) to produce alcohol and carbon dioxide. It is used to make bread and alcoholic drinks.

Figure 7.2 Dough rising to produce the spongy texture of bread.

Bread is made by mixing flour, water, yeast and sugar into a grey–white lump called dough. Inside the dough the yeast respires anaerobically and produces bubbles of carbon dioxide that make the dough rise. In a bakery the dough is cut into pieces to make loaves. Each piece

2 How does yeast make the spongy texture of the bread?

3 Why does the dough rise even more in a warm oven?

4 Why does fermentation not produce a stronger alcoholic solution than about 14%?

of dough is put into a baking tin. When bubbles of gas are heated they expand, so the tins are kept in a warm cupboard for about half an hour to allow the dough to rise even more. They are then placed in an oven for baking. The alcohol evaporates in the heat and a loaf, with a spongy texture, is produced.

Yeast and sugar are used as ingredients to make alcoholic drinks, such as wine and beer. The process of producing alcohol for drinks is called fermentation. During this process bubbles of carbon dioxide are also produced. The alcohol produced mixes with the water in the drink. Alcohol is a poison, so fermentation must be regulated to prevent it increasing to concentrations that kill the yeast. If fermentation was allowed to continue until it stopped, a solution of about 14% alcohol would be produced.

Monera

Living things in this kingdom have bodies made from only one cell that has not got a nucleus. Living things in most other kingdoms are made from many cells that contain a nucleus at some stage of their development.

The two major subgroups of the Monera are the bacteria and the blue–green algae.

Bacteria

Bacteria are single-celled organisms that range in size from 0.001 μm–0.5 μm. They can be seen using a light microscope.

Most bacteria feed on the remains of plants and animals or on animal waste. They feed by secreting a digestive juice which breaks down the food substances around them. Then they draw in the digested food substances for use in growth and reproduction.

Some bacteria can photosynthesise (see Chapter 11) or process chemicals such as hydrogen sulphide to obtain energy for life processes.

Bacteria have different body shapes as Figures 7.3 and 7.4 (pages 118 and 119) show. The body shape is used to classify bacteria. They may be round (cocci), rod-shaped (bacilli) or coiled (spirilla).

Bacteria usually reproduce by a process called fission, where each bacterium divides into two. If they have enough warmth, moisture and food, some bacteria can reproduce by fission once every 20 minutes. When conditions become dry or hot and unsuitable for feeding and breeding, some bacteria form spores. A spore has a

5 If a bacterium could divide into two every 20 minutes, how many bacteria would be produced after 8 hours?

thick wall which protects the bacterium from the hot, dry conditions. Bacteria can survive inside spores for a long time. They break out of the spores when favourable conditions return.

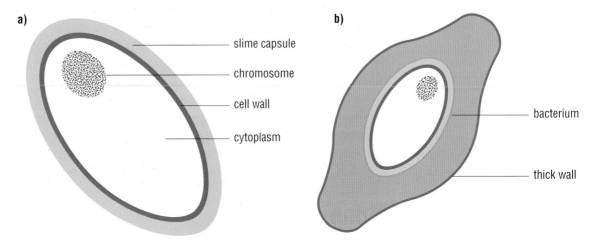

a)
- slime capsule
- chromosome
- cell wall
- cytoplasm

b)
- bacterium
- thick wall

Figure 7.3   a) A bacterium and **b)** a bacterial spore.

Growing bacteria

Bacteria can be encouraged to feed and breed by providing them with food and warmth. This is done by setting up a Petri dish containing a jelly called agar, like the one shown on page 132. In the agar are the nutrients that the bacteria require to grow.

When a bacterium settles on the agar, it feeds and breeds. Each daughter cell produced by fission also feeds and breeds, and soon a large population develops which can be seen as as waxy spot on the agar.

Great care is needed in the growing of bacteria in a Petri dish. As bacteria can settle anywhere, the Petri dish must first be sterilised. The agar must be sterilised too. Most Petri dishes are pre-sterilised by gamma irradiation.

When ready, the lid of the dish is raised slightly at one side and the agar is poured in. When the agar has set, it is ready to receive bacteria. In many studies on bacteria, the bacteria are introduced to the agar by using a wire loop. This is first sterilised by heating it in a Bunsen burner flame. It is then dipped in a liquid containing the bacteria. A film of liquid forms in the loop and the loop is moved towards the dish. The lid is again lifted slightly at one side and the loop is put into the dish and stroked from side to side across the surface

6 Why do you think that the dish and the agar are sterilised before the investigation?

7 Why is the lid only lifted slightly to add the agar and put in the wire loop?

8 The wire loop is sterilised after it has been in the petri dish. Why is this?

9 Harmful bacteria can develop if there are only conditions where anaerobic respiration can take place. Why should the lid and dish not be joined by the tape going right round the rim of the lid?

10 Body temperature is 37°C. Why are the bacteria kept below this temperature?

11 Why is the label written on the dish and not the lid?

of the agar. The lid is closed and two strips of sticky tape are used to join the lid to the dish. Petri dishes are normally incubated inverted. The underside of the dish is labelled and the dish is placed in a safe place in the laboratory and kept at room temperature or a maximum of 30°C. After a few days the colonies of bacteria can be seen through the lid or through the agar.

Some bacteria feed on the insides of living bodies where they may cause disease. Diphtheria, whooping cough, cholera, typhoid, tuberculosis and food poisoning are all caused by different kinds of bacteria.

Useful bacteria

Some kinds of bacteria are useful. Yoghurt is made by introducing bacteria to milk and making it turn sour. Vinegar is made by allowing bacteria to feed on ethanol and change it into acetic acid.

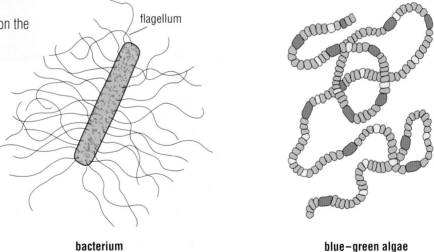

bacterium   blue–green algae

Figure 7.4   The major subgroups of the Monera.

Blue–green algae

These cells either live on their own or form threads called filaments. They may also cluster together in a jelly-like substance that they make around themselves. Blue–green algae have other coloured pigments, including red and yellow. They live in seas, oceans and lakes, where they form part of the plankton, and they make food by photosynthesis. They also grow on wet rocks at the sides of streams and rivers, at the top of rocky sea-shores, and may occur widely in the soil.

12 In what ways are members of the Monera
   a) similar to and
   b) different from members of the Protoctista?
13 Describe the way Protoctista move.

Protoctista

Many Protoctista have a body of just a single cell, although some are much bigger. They may move by making the substance inside their body flow. An *Amoeba* uses this method to form projections called pseudopodia which it uses to catch food. Other members of this kingdom may have a hair-like projection called a flagellum which they lash like a whip to move through water. Many Protoctista have bodies with small hair-like structures called cilia on their surface. They wave their cilia to-and-fro to push themselves through water.

Protoctista may either take in particles of food or make their own food by photosynthesis. Some live in the bodies of other living things and cause diseases, such as malaria and sleeping sickness.

Seaweeds are also part of the Protoctista kingdom. They have a body made from many cells.

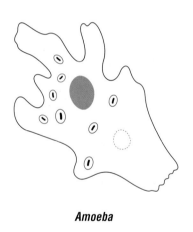

Amoeba

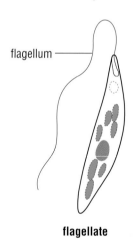

flagellum

flagellate

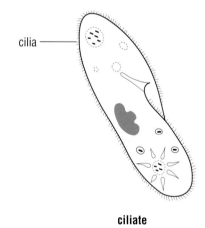

cilia

ciliate

Figure 7.5 Protoctista.

Viruses

Viruses are so small that they can only be seen with an electron microscope. Viruses do not show the characteristics of living things such as feeding, respiring or growing (see page 68). They are usually classified as non-living but they are able to reproduce if they enter a living cell. As they reproduce they destroy the cells they are in. Each kind of virus attacks certain cells in the body. For example, the cold virus attacks the cells that

14 How are viruses and bacteria different?

15 Produce a table of diseases caused by viruses and bacteria.

line the inside of the nose. The destructive action of the cold virus on the cells in the nose makes the nose run. In addition to the common cold, viruses can also cause influenza, chicken pox, measles and rabies and can lead to the development of AIDS.

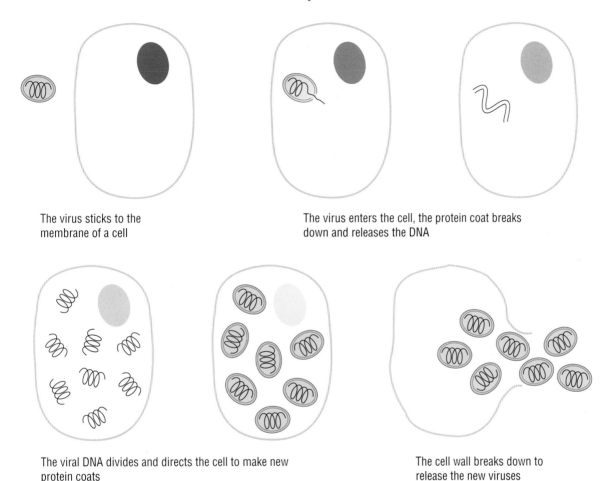

The virus sticks to the membrane of a cell

The virus enters the cell, the protein coat breaks down and releases the DNA

The viral DNA divides and directs the cell to make new protein coats

The cell wall breaks down to release the new viruses

Figure 7.6 How viruses reproduce.

Transmitting disease

Disease-causing microbes and viruses can be transmitted in several ways.

Droplets

When people cough or sneeze, they produce a cloud of tiny water droplets which can carry microbes. If other people inhale the droplets they can become infected with the microbes.

Contact

Some diseases, such as leprosy, are spread by an infected person touching an uninfected person.

Water

Water supplies may be contaminated with sewage that contains disease-causing microbes. In 1854 Doctor John Snow identified the cause of a cholera outbreak in London. A well that had previously produced pure water had become contaminated by a nearby cesspit.

Food

If food is not prepared in hygienic conditions, harmful bacteria can enter it and breed. When the food is eaten, the bacteria continue to feed and breed inside the person's body and make them ill with food poisoning.

Wounds

The skin provides a protective barrier against microbes, but if it is damaged harmful microbes can enter the tissue below, and can possibly be carried around the body in the blood.

Vectors

A vector is an animal that spreads a disease. Most vectors spread a disease when they bite someone. For example, the mosquito spreads malaria when it feeds on blood. The housefly spreads disease by contaminating food (see page 123).

Public health

In developed countries there are public health services that help to keep the population healthy. Two major features of the public health service are water treatment and waste disposal.

Water treatment

From the earliest times people have set up homes near sources of fresh water such as streams, rivers or lakes. They used the water for drinking, washing clothes, bathing and for taking away faeces and urine. In those days, though, people did not realise that there was a connection between using dirty water and diseases, such as typhoid and cholera.

For discussion
Imagine that you were making an expedition through a rainforest. What are the dangers of there being a disease transmitted to you? What preparations would you make to prevent a disease being transmitted to you?

Cleaning water to make it fit to drink and the treatment of waste water and sewage are expensive. These processes are found widely only in the wealthy developed countries. In many areas of developing countries there is not enough money available to provide these services and therefore people still suffer from diseases caused by microorganisms in the water. International aid programmes have been set up to help improve the water supplies in developing countries in order to reduce disease.

Figure 7.7   This pump is used to draw up water that has sunk into the ground and settled above water-resistant rocks.

Dangerous rubbish

Microorganisms thrive in decaying household waste such as kitchen scraps. Flies also breed in the waste and carry the microorganisms on their bodies when they leave the rubbish. If the fly lands on food left out in the kitchen microorganisms may be left behind to feed and breed. If the food is then eaten the microorganisms may cause illness. This can be prevented by storing rubbish in bins with secure lids and storing food in containers. In developed countries rubbish is collected from households and stored in tips under soil to reduce the spread of disease by flies and other pests such as rats.

Personal hygiene

During the course of the day sweat wells up from the pores in the skin. The water in the sweat evaporates to cool the body but other substances, such as urea, are left behind. Dirt sticks to the skin and dead skin cells flake off and join the dirt. Bacteria from the dirt or from the air feed and breed on these substances on the skin surface. Their activities make the skin smell and, if the skin is cut, they can enter the body and cause disease. Body odours caused by bacteria on the skin and the risk of infection are greatly reduced by a high standard of personal hygiene. This involves regular, thorough washing of skin and hair, regular changing and washing of clothes, particularly underclothes which are next to the skin, and regular tooth-brushing.

Acne

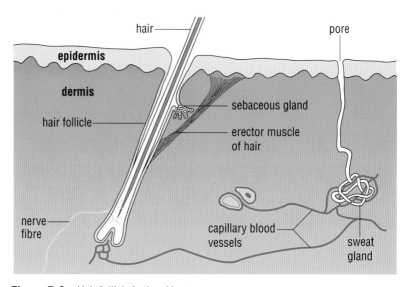

Figure 7.8 Hair follicle in the skin.

Hair grows out of tubes in the skin called follicles. There is a sebaceous gland in each follicle that secretes sebum. This substance prevents the skin becoming dry and gives it a waterproof coat. At puberty (see page 33) the body increases the production of sex hormones which may cause an increase in sebum production. The sebum can block the hair follicle and form a blackhead. Bacteria close by may breed and cause the skin to become inflamed to form a spot or pimple. Large numbers of pimples are called acne. In severe acne parts of the sebaceous glands are destroyed and cavities form in the skin. These can leave a scar.

16 Why should you clean your skin?

Microorganisms on teeth

If teeth are not cleaned regularly a sticky layer builds up on the enamel. This layer is called plaque and bacteria will settle in it. They feed on the sugar in food and make an acid. This breaks down the enamel and creates a cavity that may extend through the dentine into the pulp cavity. This has nerves running through it, so the tooth may become very sensitive, especially to hot or cold food and drink. If the cavity is not treated an abscess can form.

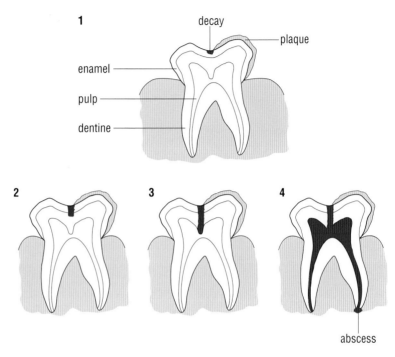

For discussion

What should you do to try to keep your teeth for life?

Figure 7.9 Stages in tooth decay.

How the body fights harmful microorganisms

The body's immune system acts to destroy harmful microorganisms. Most microorganisms are killed by white blood cells as they try to enter or soon after they enter the body through a cut, before they do any harm. These white blood cells are a type known as phagocytes. If the microorganisms do enter the blood and move round the body they come into contact with another type of white blood cell, known as lymphocytes. Microorganisms have chemicals called antigens on the surface of their bodies. Lymphocytes detect the antigens and make antibodies to attack the microorganisms and begin to destroy them. The destruction is completed by phagocytes that engulf the attacked microorganisms.

Each kind of microorganism has its own antigen which is different from any other. When the lymphocytes make an antibody to destroy an antigen it can attack only the particular kind of microorganism that makes that antigen. For example, the antibody that helps destroy the bacterium that causes whooping cough will not destroy the bacterium that causes diphtheria.

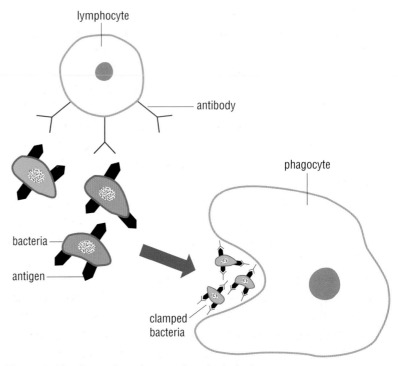

Figure 7.10 Destroying microorganisms in the body.

After a particular kind of disease-causing microorganism has been destroyed, the antibodies remain in the blood for some time. If reinfection occurs, the antibodies can destroy the invading microorganisms quickly before they build up into large numbers to cause disease. Even when the antibodies have left the blood the lymphocytes are quick to detect the antigens of the reinvading microorganism and make antibodies rapidly to begin the destruction process. The action of the lymphocytes gives the body immunity to the disease if the microorganism should reinfect it. The immunity that develops after the body has been infected with the disease-causing microorganism is called natural acquired immunity. In the past, all the immunity that a person had was built up in this way. Today, we are made immune from many microorganisms by a process called artificial immunisation.

Development of immunisation

The process of immunisation was developed in 1796 by Edward Jenner (1749–1825). Smallpox was a common disease. It was caused by a virus that infected the respiratory system then moved to the skin where it caused rashes, spots and scabs full of pus that were called pustules. Fever also developed and death often followed.

Up until that time a process called variolation had been used to try and give protection from smallpox. This had been developed in China where it was first noticed that people with a mild form of smallpox survived and did not get it again. Pus was taken from their sores and was put in the skin or up the noses of healthy people who had not had smallpox. It was thought that they too should get a mild form of the smallpox and survive. Some people did survive but many developed the virulent form and died. However, in the absence of any better process the practice of variolation was passed on to Turkey and then moved into Europe.

Edward Jenner discovered that in a village where a smallpox outbreak occurred, the milk maids remained unharmed. He found that cattle suffered from a disease similar to smallpox but its effects were milder. When the milk maids milked the cattle they became infected with cowpox. Jenner thought that it gave them protection against smallpox.

Jenner planned an experiment to test his idea by adapting the practice of variolation. He took pus from the scabs of a person who had suffered from cowpox and put it into two cuts in the arm of an 8-year-old boy. The boy developed cowpox but quickly recovered. Seven weeks later Jenner took some pus from the scab of a smallpox patient and put some of the pus in the cuts. The boy did not get smallpox. Jenner called this process vaccination after the Latin word for cowpox, 'vaccinia'. Soon vaccination was widely practised. Today no-one suffers from smallpox.

Fifty years after Jenner's experiment, Louis Pasteur (1822–1895) discovered a way of making sheep immune to anthrax. There is no mild form of anthrax so Pasteur looked at ways of weakening the microorganisms that produced the fatal disease. He discovered that if the microorganisms were heated then given to sheep they produced a mild form of the disease. When sheep recovered he gave them the normal microorganisms but they did not develop the fatal form of the disease. Weakened or attenuated microorganisms are used in immunisation schemes today.

1 What was the main problem with variolation?
2 What could have happened if Jenner had not been right?

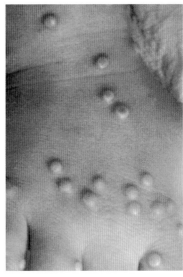

Figure A   The hand of a smallpox victim.

Artificial acquired immunity

The body can be made immune from infection without it having to suffer the full effects of the disease. This is done by vaccination. The body is injected with material that stimulates the lymphocytes to produce antibodies ready to attack those particular microorganisms if they infect the body in the future. The material that is injected may be one of the following.

- Living microorganisms that have been weakened so that they cannot cause the disease before the lymphocytes produce antibodies to destroy them. The microorganisms that produce poliomyelitis, smallpox and tuberculosis are used for vaccination in this form.
- Dead microorganisms. Their bodies still have the antigens that stimulate the lymphocytes to produce antibodies. The microorganisms that produce influenza, typhoid and whooping cough are used for vaccination in this form.
- Poisonous toxins made by bacteria that have been treated to make them harmless. The toxins that cause diphtheria and tetanus are used for vaccination in this form.

While the immunity given by some vaccines can last a lifetime, other vaccines, such as those for cholera and typhoid, only give immunity for a certain length of time. Extra vaccinations, called boosters, are needed to keep the body immune from these diseases.

Passive immunity

The fetus developing in the womb does not make antibodies but receives some from the mother's blood through the placenta (see page 43). After the baby is born it may take in more antibodies through its mother's milk. Immunity built up in this way is called passive immunity. Soon the baby starts to make its own antibodies to build up immunity.

17 What is the difference between
   a) an antibody and an antigen,
   b) a lymphocyte and a phagocyte?
18 How does the body fight reinfection
   a) shortly after it has recovered from the disease,
   b) a long time after it has recovered from the disease?
19 What are the benefits of artificial immunity?
20 How does society benefit from mass vaccinations against polio?

Building up protection

In the United Kingdom each young person is helped to build up immunity by receiving the series of vaccinations listed in Table 7.1 in their early life.

Immunisation against German measles (rubella) is provided to prevent the occurrence of rubella during pregnancy later in life. If a pregnant woman, who has not been immunised against German measles, is infected with the virus, it could also damage the tissues of the developing fetus. When born the baby may suffer deafness, mental handicap, disorders of the eyes and damage to the heart and liver.

Table 7.1 The immunisation schedule for children and young people in the United Kingdom.

Age	Immunisation
2 months	Diphtheria, tetanus, whooping cough, HIB, polio
3 months	Diphtheria, tetanus, whooping cough, HIB, polio
4 months	Diphtheria, tetanus, whooping cough, HIB, polio
12–18 months	Measles, mumps, rubella
4 years	Diphtheria, tetanus, whooping cough booster, measles, mumps, rubella
13–14 years	Tuberculosis (BCG)
15 years	Diphtheria, tetanus, polio

HIB = immunisation against *Haemophilus influenzae* type B disease, which can cause meningitis.

The letters BCG stand for Bacille Calmette-Guérin. It is named after the two French bacteriologists Calmette and Guérin who first produced the vaccine.

Operations today and yesterday

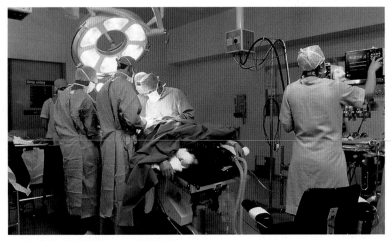

Figure A An operation in progress.

In this operation the surgeon cut open the patient's body to work on an organ inside. At the end of the operation the cut was stitched up and the patient returned to the ward for a few days before going home to make a full recovery. If microorganisms had entered the body during the operation they could have caused infections. They are prevented from doing so in the following ways. The air is filtered before it enters the operating theatre and the smooth, easily cleaned surfaces are washed with disinfectant. The instruments used in the operation and the caps, gowns and gloves of the people taking part are sterilised to kill any microorganisms that may have settled on them.

Joseph Lister (1827–1912) was a surgeon. In his day surgeons could perform successful operations but many of the patients died later because their wounds became infected and turned septic. When Lister heard about Pasteur's 'germ theory' he believed that microorganisms could be entering the patient's body during operations and causing the infections to develop. He decided to use a carbolic acid spray during his operations to kill any microorganisms that might be present. He found that his patients did not develop infections. Carbolic acid was the first antiseptic substance. Many more have been developed since Lister's time.

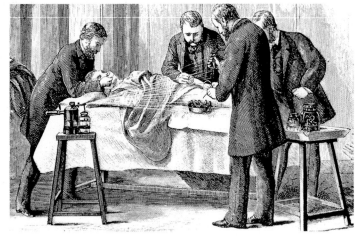

Figure B Lister's carbolic spray in action.

1 There is usually only one surgeon in a modern operating theatre. What do you think are the jobs of the other people in the room?

2 What differences can you see between the photograph of the modern operating theatre and the picture from Lister's time (Figure B)?

3 What changes could the surgeons in Lister's day have made to help keep microorganisms from infecting the patient?

4 How did antiseptics get their name?

5 What do you do to prevent small wounds from becoming septic?

Fighting microorganisms with medicines

Bacteria can be destroyed with an antibiotic. Antibiotics are chemicals that are swallowed or injected to fight microorganisms inside the body. The antibiotic may stop the bacteria from making cell walls or it may affect the life processes taking place inside them. Two well known antibiotics are penicillin and tetracycline.

Viruses cannot be destroyed by antibiotics but some can be destroyed by antiviral drugs. A virus needs substances from the cell it has infected in order to reproduce. An antiviral drug stops the virus reaching these substances.

Antiseptics are chemicals used to attack micro-organisms on damaged skin and in the lining of the mouth. They are not swallowed. They are used as creams and mouthwashes.

Athlete's foot is a skin disease caused by a fungus. The fungus feeds on the damp skin between the toes of poorly dried feet. The disease can be cured by treating the feet with powder and cream containing a fungicide which kills the fungus.

21 What are the differences between an antibiotic and an antiviral drug?

The first antibiotic

Bacteria and fungi are two kinds of microorganisms which can be grown on plates of agar jelly that are sealed inside a petri dish. Agar jelly contains the nutrients that bacteria and fungi require.

In 1928 Alexander Fleming (1881–1955) was working in a laboratory when he noticed that a plate that had been set aside to grow colonies of a certain bacterium also had a green fungus growing on it. At first sight it looked as if the plate had been spoiled, but when Fleming looked again he saw that the bacterial colonies near to where the fungus was growing had been destroyed. He reasoned that there was a substance in the jelly, which had come from the fungus, that killed the bacteria. The fungus was called penicillium which means 'little brush'. It was given this name because under the microscope parts of the fungus look like little brushes. When Fleming extracted the substance from the fungus he called it penicillin.

Fleming tested penicillin on a range of bacteria and found that some were killed but others were not. He then tested penicillin on human white blood cells and found that they were not destroyed by concentrations that killed the bacteria. Fleming did not investigate penicillin further to discover its chemical structure, and when he published his work other scientists were not interested in it.

1 Why can bacteria and fungi grow well on agar?

2 Why was Fleming's agar plate thought to have been spoiled?

3 Why did Fleming believe the fungus made a substance that killed bacteria?

4 Look at Figure A (page 132). Which disc has the strongest antibiotic? Explain your answer.

5 How did the results of Fleming's experiment on human white blood cells suggest that penicillin could be useful?

(continued)

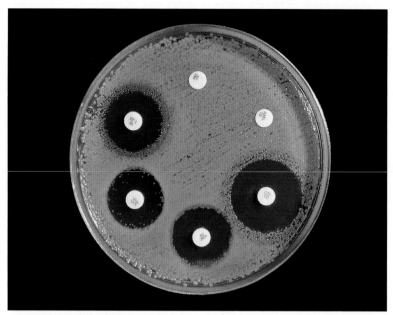

Figure A Discs of antibiotics on a plate originally covered with a suspension of a bacterium called *Escherichia coli*. Plates of agar jelly containing bacterial colonies are still used to test antibiotics today.

When the United Kingdom was preparing for World War II, a research programme was set up to find a substance that could be used to kill bacteria in soldiers' wounds. Ernst Chain (1906–1979) was working on the programme and looked up some work done by Fleming in 1922. In this work Fleming had discovered lysozyme, which is a chemical that kills bacteria and is found in tear drops. Chain also read about Fleming's discovery of penicillin. He told a scientist he was working with called Howard Florey (1898–1968) about it. Florey and Chain decided that penicillin could be the substance that was needed and began studying it in more detail. Their work led to penicillin being used to treat wounded soldiers in the war. In 1945 Fleming, Florey and Chain were awarded the Nobel Prize for Medicine.

Further investigations on penicillin continued during the war. They were made by Dorothy Hodgkin (1910–1994) who studied the crystal structure of chemicals by firing beams of X-rays at them. When the X-rays strike the chemical molecules they move off in new directions. These directions can be found by using a photographic film set-up around the chemical. The way to work out the structure of a molecule from the X-ray paths is very complicated and when Hodgkin used it to investigate penicillin she used a computer. This was the first time that a computer had been used in a biochemistry experiment. When Hodgkin had discovered the structure of penicillin she went on to investigate the structures of vitamin B_{12} and insulin. In 1964 she was awarded the Nobel Prize for Chemistry. The following year she was awarded the Order of Merit. The last woman to receive this award had been Florence Nightingale.

6 How did a bacteria-killing substance affect wounded soldiers?

7 How do you think penicillin affected the death rate in army hospitals?

Figure B Dorothy Hodgkin.

◆ SUMMARY ◆

♦ Microbes belong to three kingdoms of living things – fungi, Monera and Protoctista (*see page 115*).

♦ Yeast is a useful fungus (*see page 116*).

♦ There are special techniques for growing bacteria (*see page 118*).

♦ Bacteria may be harmless or may cause disease (*see page 119*).

♦ Viruses are usually classified as non-living (*see page 120*).

♦ Diseases can be transmitted in several ways (*see pages 121–122*).

♦ Water treatment is important for community health (*see page 122*).

♦ Skin hygiene and dental care are important components of personal hygiene (*see pages 123–125*).

♦ The body's immune system naturally helps destroy bacteria and viruses (*see pages 125–126*).

♦ Immunity can be acquired artificially (*see page 128*).

♦ Medicines can be used to destroy microbes (*see page 131*).

End of chapter questions

1 Which organisms
  a) have cell walls made of chitin,
  b) can make food by photosynthesis,
  c) have round, rod-like or spiral shapes?
2 Explain why antibiotics cannot be used to cure the common cold.
3 How are microbes useful?

8 Ecological relationships

There are five kingdoms of living things. They are the animal kingdom (see page 71), the fungi kingdom (see page 115), the Monera kingdom (see page 117), the Protoctista kingdom (see page 120) and the plant kingdom (below).

Plants

Plants are made from a large number of cells. Each cell has a cell wall made of cellulose. Plants make their own food by photosynthesis. The plant kingdom is divided into four groups.

Liverworts and mosses

Liverworts are small plants that do not have true roots, stems or leaves. They grow in damp places near streams and ponds.

Mosses have stems and leaves but they do not have proper roots. Moss plants are usually found growing together in many different habitats, from dry walls to damp soil.

Both liverworts and mosses reproduce by producing spores. They make the spores in a capsule that is raised into the air. When the capsule opens the spores are carried away by air currents.

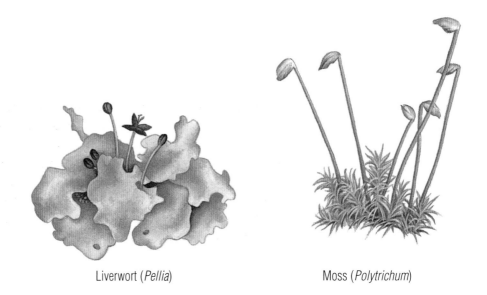

Liverwort (*Pellia*)   Moss (*Polytrichum*)

Figure 8.1 A liverwort and a moss plant.

Ferns and horsetails

These plants have true roots and stems and reproduce by making spores. In ferns the spores are made in structures called sporangia on the underside of the large feather-like leaves called fronds. Horsetails produce a cone-like structure at the tip of the shoot that makes and releases spores. In both kinds of plant the spores are carried away by air currents.

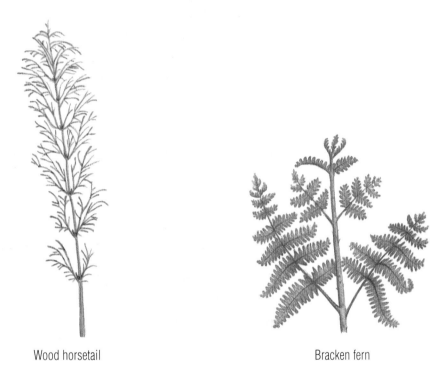

Wood horsetail   Bracken fern

Figure 8.2   A horsetail and a fern.

Conifers

A conifer has roots, a woody stem and needle-like leaves. Most conifers lose and replace their leaves all year round, so they are called evergreen. Almost all conifers reproduce by making seeds that develop in cones. When the seeds are ready to be dispersed the cones open and the seeds fall out. Each seed has a wing that prevents the seed falling quickly and allows it to be blown away by the wind.

Figure 8.3　Male and female cones on a conifer.

Flowering plants

A flowering plant has a root, stem and leaves. In some plants the stem is woody. All of these plants reproduce by flowering and making seeds (see Chapter 2).

Figure 8.4　Grass, bluebells and these trees are all flowering plants.

Investigating a habitat

A habitat, such as wood or a pond, is a home for a range of living things. When ecologists study a habitat they need to collect data about it. This not only provides information about the organisms that live there, but the data can also be stored and used to monitor the habitat in the following way. At a later date another survey can be made and the data obtained then can be compared with the previous data. This shows how the populations of species have fared in the time between the surveys. Some may have increased, others decreased, and some stayed the same. By comparing data in this way the effects of events close by, such as a change in land use or the release of pollutants, can be carefully studied for signs of environmental damage.

Recording the plant life in a habitat

When a habitat is chosen for study a map is made in which the habitat boundaries and major features, such as roads or cliffs, are recorded. The main kinds of plants growing in the habitat are identified and the way they are distributed in the habitat is recorded on the map.

A more detailed study of the way the plants are distributed is made by using a quadrat and by making a transect.

Figure 8.5 Ecologists mapping a habitat.

Using a quadrat

A quadrat is a square frame. It is placed over an area of ground and the plants inside the frame are recorded. If there are only a few plants in the quadrat, as may occur if it is placed on waste ground, the positions of the individual plants can be recorded in a diagram of the quadrat. If the quadrat is placed in an area with a large number of plants covering the ground, such as in a lawn, the area occupied by each type of plant is estimated. For example, a quadrat on a lawn may show the plants to be 90% grass, 7% daisy and 3% dandelion.

When using a quadrat the area of ground should not be chosen carefully. A carefully selected area might not give a fair record of the plant life in the habitat but may support an idea that the ecologists have worked out beforehand. To make the test fair the quadrat is thrown over the shoulder so that it will land at random. The plants inside it are then recorded. This method is repeated a number of times and the results of the random samples are used to build up a record of how the plants are distributed. An estimate of how many of each kind there are in the area can then be made.

1 How could you use a quadrat to see how the plants change in a particular area over a year?

Figure 8.6 Ecologists using a quadrat.

Making a transect

If there is a feature such as a bank, a footpath or a hedge in the habitat, the way it affects plant life is investigated by using a line transect. The position of the transect is chosen carefully so that it cuts across the feature being examined.

The transect is made by stretching a length of rope along the line to be examined and recording the plants growing at certain intervals (called stations) along the rope. When plants are being recorded along a transect, abiotic factors such as temperature or dampness of the soil may also be recorded to see if there is a pattern between the way the plants are distributed and the varying abiotic factors.

Figure 8.7 Ecologist making a transect.

2 How useful do you think quadrats and line transects are for recording the positions of animals?

3 Which method – quadrat or transect – would be more useful for investigating the plants growing around the water's edge of a pond? Explain your answer.

4 Look at the results from a line transect shown in Table 8.1.

Table 8.1

Station	1	2	3	4	5	6	7	8	9	10
Soil condition	W	D	W	W	D	Dr	Dr	Dr	D	Dr
Plant present	A	A	B	A	B	C	C	C	B	C

W = very wet soil; D = damp soil; Dr = dry soil

What can you say about the plants in this habitat from the information in these results?

Collecting small animals

Different species of small animals live in different parts of a habitat. In a land habitat they can be found in the soil, on the soil surface and leaf litter, among the leaf blades and flower stalks of herbaceous plants, and on the branches, twigs and leaves of woody plants. They can be collected from each of these regions using special techniques.

Collecting from soil and leaf litter

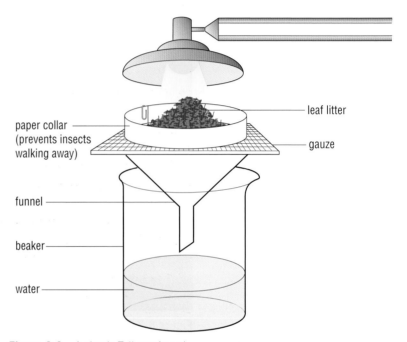

Figure 8.8   A simple Tullgren funnel.

A Tullgren funnel is used to collect small animals from a sample of soil or leaf litter. The sample is placed on a gauze above the funnel and a beaker of water is placed below the funnel. The lamp is lowered over the sample and switched on. The heat from the lamp dries the soil or leaf litter and the animals move downwards to the more moist regions below. Finally, the animals move out of the sample and into the funnel. The sides of the funnel are smooth so the animals cannot grip onto them and they fall into the water.

Pitfall trap

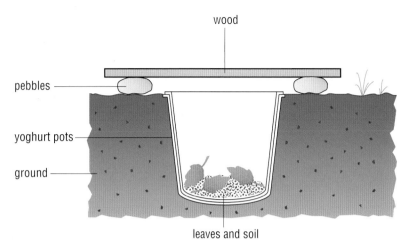

Figure 8.9 A pitfall trap.

The pitfall trap is used to collect small animals that move over the surface of the ground. A hole is dug in the soil to hold two containers, such as yoghurt pots, arranged one inside the other. The containers are placed in the hole, and the gap around them up to the rim of the outer container is filled in with soil. A few small leaves are placed in the bottom of the container and four pebbles are placed in a square around the top of the trap. A piece of wood is put over the trap, resting on the pebbles. The wood makes a roof to keep the rain out and hides the container from predators. When a small animal falls in it cannot climb the smooth walls of the inner container and remains in the trap, hiding under the leaves until the trap is emptied. Traps must be emptied after a few hours and those set in the evening must be emptied the following morning. The animals are collected by removing the cover, taking up the inner container and emptying it into a white enamel dish. The animals can be seen clearly against the white background and identified.

Sweep net

The sweep net is used to collect small animals from the leaves and flower stems of herbaceous plants, especially grasses. The lower edge of the net should be held slightly forward of the upper edge to scoop up the animals as the net is swept through the plants. After one or two sweeps the mouth of the net should be closed by hand and the contents emptied into a large plastic jar where the animals can be identified.

5 What is the advantage of using an outer and an inner container instead of just one container for the pitfall trap?

6 Why are large leaves not used inside the trap?

Figure 8.10 Using a sweep net.

Sheet and beater

Small animals in a bush or tree can be collected by setting a sheet below the branches and then shaking or beating the branches with a stick. The vibrations dislodge the animals, which then fall onto the sheet. The smallest animals can be collected in a pooter (Figure 8.11).

Pooter

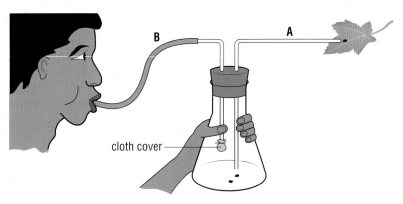

cloth cover

Figure 8.11 Using a pooter.

7 What is the purpose of the cloth on tube B of the pooter?

Tube A of the pooter is placed close to the animal and air is sucked out of tube B. This creates low air pressure in the pooter so that air rushes in through tube A carrying the small animal with it.

Collecting pond animals

Pond animals may be collected from the bottom of the pond, the water plants around the edges or the open water just below the surface.

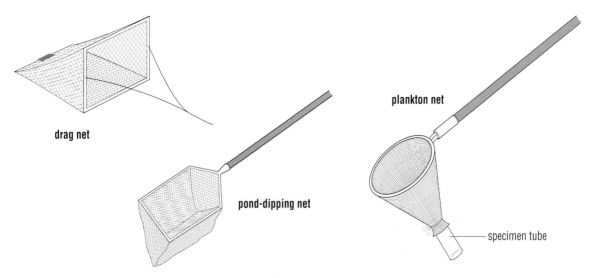

Figure 8.12 Three types of pond net.

A drag net is used to collect animals from the bottom of the pond. The net is dragged across the bottom of the pond. As it moves along it scoops up animals living on the surface of the mud. The pond-dipping net is used to sweep through vegetation around the edge of the pond to collect animals living on the leaves and stems. A plankton net is pulled through the open water to collect small animals swimming there.

The drag and pond-dipping nets are emptied into white enamel dishes so that the animals can be identified and studied. The dishes should be set up out of full sunlight so that the water does not get hot, and they should be emptied back into the pond as soon as the investigation is complete. The plankton net has a small bottle that is examined with a hand lens. Samples are then taken for microscope examination.

Studying animals more closely

After a habitat has been surveyed and the different species of plants and animals identified and recorded, further studies about their lives can be made. These involve making observations. If the species to be studied is a plant, a record of its growth, flowering and seed dispersal may be made. If the species is an animal, its behaviour may be observed over different times of the day and throughout the year. Birds are relatively easy to study because most species do not hide away. Small animals hide away but they can be kept in containers in the laboratory for their activities to be observed. Any

8 Why do you think bird-watching is a popular hobby?

9 Why should laboratory observations be compared with observations of animals in their habitat?

observations made in the laboratory must be compared to observations of animals living in their habitat before any firm conclusions are drawn, and the laboratory animals should be returned to their habitat as soon as all the observations on them are complete.

Animals in the laboratory

Small animals can be studied in the laboratory or animal house by setting up a habitat that is similar to their own.

Snails

Snails can be kept in an aquarium tank. The floor of the tank should be covered with a mixture of damp soil and peat. Large stones may be placed in the mixture for the snails to climb on and hide under.

Figure 8.13 A snail tank.

Woodlice

Woodlice can also be kept in an aquarium tank if the floor is covered with soil and a layer of damp moss and pieces of bark.

Figure 8.14 A woodlouse tank.

10 The woodlice were also observed in damp conditions but not in dry ones. How would you set up the shallow tray to test this observation?

11 When checking the behaviour of the woodlice in their habitat some were found under a log one day and under a stone about a metre away the next day. When do you think they moved? Explain your answer.

Investigating behaviour

Experiments can be devised to investigate the way the animals behave after observing them in the tanks. For example, it may be noticed that the woodlice are found under the bark in the daytime. This observation may lead to the idea that woodlice do not like the light. This idea can be tested by putting the woodlice in a shallow tray, part of which is uncovered and in the light and part of which is covered and in the dark. The woodlice should be placed in the centre of the tray and left for a few minutes before recording where they have settled to rest.

Ecological pyramids

When a survey of a habitat is complete, ecologists can examine the relationships between the different species. They may find for example that deer depend on ferns for shelter in a wood, and that some birds use moss to line their nests. The major relationship between organisms in a habitat is the relationship through feeding. It is this relationship which interests ecologists (scientists who study ecosystems).

By studying the diets of animals in a habitat, ecologists can work out food chains (see page 60) and ecological pyramids.

Pyramid of numbers

The simplest type of ecological pyramid is the pyramid of numbers. The number of each species in the food chain in a habitat is estimated. The number of plants may be estimated using a quadrat (see page 137). The number of small animals may be estimated by using traps, nets and beating branches (see pages 141–143). The number of larger animals, such as birds, may be found by observing and counting.

An ecological pyramid is divided into tiers. There is one tier for each species in the food chain. The bottom tier is used to display information about the plant species or producer. The second tier is used for the primary consumer and the tiers above are used for other consumers in the food chain. The size of the tier represents the number of each species in the habitat. If the food chain:

grass → rabbit → fox

is represented as a pyramid of numbers, it will take the form shown in Figure 8.15 overleaf.

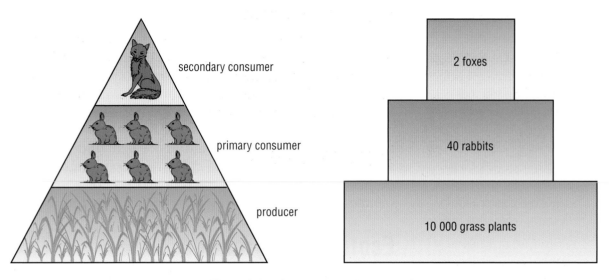

Figure 8.15 Food pyramid of numbers of grass plants, rabbits and foxes.

12 What would happen to the number of rabbits and grass plants if the number of foxes
  a) increased and
  b) decreased?

13 What would happen to the number of grass plants and foxes if the number of rabbits
  a) increased and
  b) decreased?

14 Why do the two food chains considered here produce different pyramids of numbers?

15 Why do you think there are usually more organisms at the bottom of a food chain?

Not all pyramids of numbers are widest at the base. For example, a tree creeper is a small brown bird with a narrow beak that feeds on insects that in turn feed on an oak tree. The food chain of this feeding relationship is:

oak tree → insects → tree creeper

When the food chain is studied further and a pyramid of numbers is displayed it appears as shown in Figure 8.16.

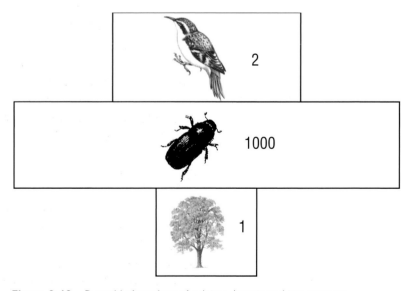

Figure 8.16 Pyramid of numbers of oak tree, insects and tree creepers.

16 Living things need water in their bodies to survive. What happens to the living things used to work out a pyramid of biomass? Explain your answer.

17 If you drew a pyramid of biomass for the food chain

oak tree → insects → tree creeper

what do you think it would look like? How would it compare to the pyramid of numbers? Explain any differences that you would see.

Pyramid of biomass

The amount of matter in a body is found by drying it to remove all water then weighing it. This amount of matter is called the biomass. Ecologists find measuring biomass useful as it tells them how much matter is locked up in each species of a food chain.

Decomposers

Figure 8.17 These Malaysian termites are feeding on leaf litter.

Not only do the living bodies of each species provide food for others but their dead bodies and waste are food too. The dead bodies of plants and animals are food for fungi, bacteria and small invertebrates that live in the soil and leaf litter. These organisms are called decomposers. When they have finished feeding, the bodies of plants and animals become reduced to the substances from which they were made. For example, the carbohydrates in a plant are broken down to carbon dioxide and water as the decomposers respire. Other substances are released from the plant's body as minerals and return to the soil. Decomposers are recyclers. They recycle the substances from which living things are made so that they can be used again.

18 Why are decomposers important? How do they affect you?

Ecosystems

Decomposers form one of the links between the living things in a community and the non-living environment. Green plants form the second link. When a community of living things, such as those that make up a wood, interact with the non-living environment – the decomposers releasing minerals, carbon dioxide and water into the environment and then plants taking them in again – the living and non-living parts form an ecological system or ecosystem. An ecosystem can be quite small, such as a pond, or as large as a lake or a forest.

19 How might studying ecosystems help to conserve endangered species?

Working out how everything interacts is very complicated but is essential to ecologists if they are to understand how each species in the ecosystem survives and how it affects other species and the non-living part of the ecosystem. Figure 8.18 shows how the living and non-living parts of a very simple ecosystem react together.

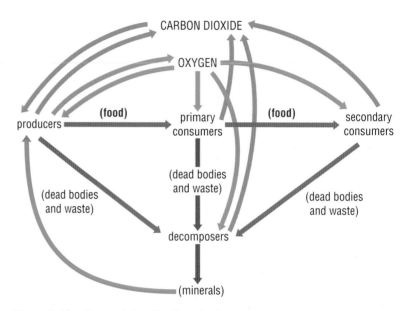

Figure 8.18   Some relationships in a simple ecosystem.

20 An aquarium tank set up with pond life is an ecosystem. See Figure 8.19.

a) Which are the producers and which are the consumers?

b) Construct some food chains that might occur in the tank.

c) Where are the decomposers?

d) Give examples of the ways the living things react to their non-living environment.

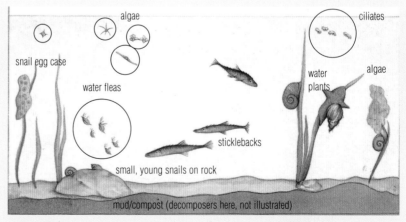

Figure 8.19   An aquarium with pond life. The circled organisms are greatly magnified.

Indicators of pollution

Some living things are very sensitive to pollution and therefore can be used as biological indicators of pollution.

Lichens are sensitive to air pollution. Where the air is very badly polluted no lichens grow but a bright green Protoctista called *Pleurococcus* may form a coating on trees. Crusty lichens, some species of which are yellow, can grow in air where there is some pollution. Leafy lichens grow where the air has only a little pollution. Bushy lichens can only grow in unpolluted air.

Some freshwater invertebrates can be used to estimate the amount of pollution in streams and rivers. If the water is very badly polluted there is no freshwater life but if the water is quite badly polluted rat-tailed maggots may be present. Bloodworms can live in less badly polluted water and freshwater shrimps can live in water that has only small amounts of pollution. Stonefly nymphs can only live in unpolluted water.

high pollution

no lichens – only *Pleurococcus*

crusty lichen

increasing pollution

leafy lichen

clean air   bushy lichen

Figure A   Lichens.

1 How polluted is the habitat if:
   a) the trees have Pleurococcus and crusty yellow lichens on them?
   b) bushy, leafy and crusty lichens are found in a habitat?
   c) freshwater shrimps and bloodworms are found in a stream?

2 Four places were examined in succession along a river and the animals found there were recorded. Here are the results.

Station A) Stonefly nymphs, mayfly nymphs, freshwater shrimp, caddis-fly larvae.

Station B) Rat-tailed maggots, sludge worms.

Station C) Sludge worms, bloodworms, waterlouse.

Station D) Freshwater shrimps, waterlouse.

What do the results show? Explain your answer.

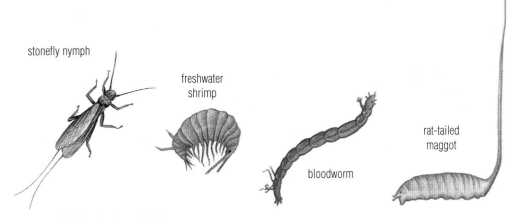

stonefly nymph

freshwater shrimp

bloodworm

rat-tailed maggot

Figure B   Freshwater invertebrates.

How populations change

If an area of ground is cleared of vegetation it will soon be colonised by new plants and animals. The following is an account of how an area of soil could be colonised. The colonisation is much simpler than occurs naturally so that the ways the plants and animals interact can be seen more clearly.

A seed lands in the centre of a soil patch and germinates. The plant is an ephemeral (has a very short life cycle) and is soon fully grown and producing flowers and seeds. The seeds are scattered over the whole area. They all germinate and grow so the population of the ephemeral plants increases.

There are perennial plants (which flower and continue to grow for several years) outside the area of cleared soil that have stems with broad leaves that cover the ground. As the population of the ephemeral plants increases, the perennial plants grow into the area of cleared soil. The two kinds of plants compete for light, water and minerals in the soil. As the numbers of both plants increase, the competition between them also increases. The perennials compete more successfully than the ephemeral plants for the resources in the habitat and produce more offspring. The broad leaves of the perennial plants cover the soil and prevent seeds from landing there and germinating. The leaves may also grow over the young ephemeral seedlings. In time the ephemeral plants that are producing seed will die and the stems and broad leaves of the perennials could cover them too.

Herbivorous insects land on some of the perennial plants and start to feed on their leaves. They feed and breed and as their numbers increase they spread out over other perennial plants in the area. The population of the perennial plants in the area begins to fall and the population of ephemerals, which are not eaten by the insects, begins to rise.

A few carnivorous insects land in the centre of the patch. They clamber about on the plants and feed on the herbivorous insects. The well fed carnivorous insects breed and their population increases. The population of the herbivorous insects starts to fall.

21 In what ways do the two kinds of plants compete for the resources?

22 If the herbivorous insects had not arrived what do you think would have happened to the two populations of plants?

23 How did the arrival of the herbivorous insects affect your prediction in question 22? Explain your answer.

24 What effect do the carnivorous insects have on the population of
a) herbivorous insects and
b) ephemeral plants?

25 How may the population of herbivorous insects change over a period of time?

26 Draw a freehand graph to show the change in size of the population of
a) herbivorous insects and
b) carnivorous insects over time.

Predicting changes in populations

The changing size of the human population can be predicted by comparing birth rates with death rates. The birth rate is the number of babies born per 1000 people in the population in a year. The death rate is the number of people dying per 1000 people in the population in a year.

If the birth rate is greater than the death rate the population will increase in size. If the death rate is greater than the birth rate the population will decrease in size. If the birth rate and the death rate are the same the population will remain unchanged.

Birth rates, death rates and conservation

Many endangered mammal species have been reduced to a small world population by hunting. The animals have been killed more quickly than they can reproduce. If the death rate exceeds the birth rate the mammal species is set on a course for extinction. Many mammals are now threatened with this course.

They can be helped by raising their birth rate and reducing their death rate. Zoos can help increase the size of the world population of some endangered animal species. They do this by increasing the birth rate by making all the adult animals in their care healthy enough to breed and by providing extra care in the rearing of the young. Zoos also reduce the death rate by protecting the animals from predation. In many countries reserves have been set up in which endangered animals live naturally but are protected from hunting by humans. This reduces the death rate, which in turn increases the birth rate as more animals survive to reach maturity and breed.

For discussion

Large mammals need large areas of natural habitat to support a large population. With the increasing human population why is it difficult to conserve these large areas? Explain your answer.

Figure 8.20 A Java rhinoceros which is threatened with extinction.

Habitat destruction

The first humans did not destroy habitats. They hunted animals and gathered plants to eat in the same way that a few groups of people in the rainforests still do today. When people discovered how to raise crops and farm animals the size of the human population began to rise slightly as there was more food to eat.

In the past few hundred years the size of the human population has grown rapidly as people become healthier, live longer and produce more children. Towns and cities have been set up and habitats destroyed to provide the space for them and the farm land needed to support them. Today, with a human population of more than 6 billion, habitats are being destroyed every moment of the day to provide extra room. Tropical rainforests are being destroyed at the rate of an area the size of a football pitch every second to make room for farms, roads and towns.

Woodlands are being destroyed to make space for roads, houses and factories. Hedgerows have been removed to make larger fields for growing crops. Freshwater habitats such as lakes and rivers have been polluted with sewage and industrial wastes. These pollutants have also been released into the sea where marine habitats are also at risk from oil pollution from shipwrecked tankers.

Figure A   Habitat destruction caused by building a road.

Figure B   An oil-polluted beach.

(continued)

Power stations provide electricity. This makes our lives easier but acid rain caused by power station smoke has destroyed forests and polluted lakes and rivers. The carbon dioxide produced by burning coal and oil in power stations adds to the greenhouse effect and leads to global warming. Carbon dioxide in the atmosphere acts like glass in a greenhouse. It lets heat energy from the Sun pass through it to the Earth's surface but it does not let heat from the Earth's surface pass through the atmosphere into space. The heat is trapped in the atmosphere and causes the temperature of the atmosphere to rise. Without this rise in temperature the Earth would be too cold for living things to survive.

Adding more and more carbon dioxide to the atmosphere leads to an increased temperature. This is called global warming and leads to melting of the polar ice, raising of the sea-level and changes in climate in all places on the planet.

Power stations that use nuclear energy do not produce gases that add to the greenhouse effect and global warming. However, they do present the risk of the leak of radioactive materials. The radiation from these materials can destroy any kind of habitat.

Fertilisers increase the amount of food that can be grown and so make it cheaper (see page 211). The use of too much fertiliser has meant that some of it drains into freshwater habitats and can cause an overgrowth of algae. When the algae die back large numbers of bacteria develop to decompose them. The bacteria use up so much oxygen from the water that other forms of water life suffocate and die.

More food, more living space, more electrical energy, more fuel for vehicles and more materials to use in a wide range of ways have all improved the quality of life for millions of people, but to provide all of these things habitats have had to be removed. In the past, when the human population was small, the scale of habitat destruction was also small. Today, with a huge human population, the scale of habitat destruction is vast. Many people feel that more care should be taken in balancing the needs of humans with the destruction of the remaining natural habitats. In many countries there are laws that protect some habitats from destruction and therefore any changes to the remaining habitats have to be carefully planned.

1 Why do habitats have to be cleared?

2 How have habitats been destroyed by other forms of human activity?

3 Try to imagine the lifestyle of the early humans who hunted animals and gathered plants. Compare this lifestyle with your own. What would you change in your own lifestyle to prevent habitat destruction?

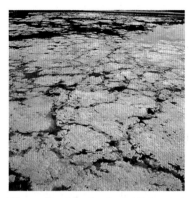

Figure C A freshwater habitat polluted by fertiliser.

(continued)

4 Figure D shows the position of two coastal towns.
A and B are two towns that have a fishing industry. Due to over-fishing the industry has declined. There are large numbers of people now unemployed in both towns and many are thinking of moving or travelling to the cities to find work. It is proposed to build an oil refinery near one of the towns. This will bring employment for the people in the form of building and running the refinery, and in the factories that may be set up to use its products. Land will be needed for the refinery and for the port where the oil tankers will dock. Land will also be needed for factories and perhaps housing estates if more people come to live in the town.

Figure D Map showing positions of towns A and B.

a) What else may land be needed for if more people come to the town?

b) What habitats may be affected by the building of the oil refinery? Explain your answers.

For discussion

What are the advantages and disadvantages of choosing to build the refinery near town A or B?

What are the major issues involved in deciding where the refinery is to be built?

How would you balance these issues to decide which town is best suited for the refinery and the port?

♦ SUMMARY ♦

♦ In the plant kingdom are mosses, liverworts, ferns, horsetails, conifers and flowering plants (*see pages 134–136*).
♦ Plant distribution can be examined using a quadrat and by making a transect (*see pages 137–139*).
♦ The Tullgren funnel, pitfall trap, sweep net, sheet and beater can be used to investigate small animals in land habitats (*see pages 140–142*).
♦ A small range of nets may be used to collect small animals from aquatic habitats (*see page 143*).
♦ Some small animals can be kept in the laboratory and their behaviour can be studied by harmless experiments (*see pages 144–145*).
♦ A pyramid of numbers shows the numbers of each species at each link in the food chain (*see pages 145–146*).
♦ Decomposers break down the dead bodies and wastes of living things into simple substances (*see page 147*).
♦ The living and non-living parts of a habitat form an ecosystem (*see pages 147–148*).
♦ Populations change due to competition, predation and adaptation (*see page 150*).
♦ Birth rates and death rates are important in predicting changes in populations (*see page 151*).

End of chapter questions

1 How would you set about investigating a habitat such as a hedgerow?
2 What surfaces do river limpets prefer? The river limpet lives in fast-flowing streams and rivers. There are many different surfaces on the bottom of a stream or river. This experiment was devised to test the results of observations in streams and rivers. The experiment also tested limpets of different sizes to see if larger ones had different preferences to smaller ones.

A 9 cm crystallising dish was divided into four sections called sectors. Each sector had one type of material, either sand, grit, stone or glass. The sectors were covered with water. Twelve limpets were used in each trial and there were two trials for each size class. New limpets were used for each trial. During each trial the number of limpets in each section was recorded after 30 minutes and 60 minutes. The results are displayed in Table 8.2.

Table 8.2

Size class/ mm	30 minutes						60 minutes					
	Sand	Grit	Stone	Glass	Total Rough	Smooth	Sand	Grit	Stone	Glass	Total Rough	Smooth
3–4	5	1	9	9	6	18	3	0	15	6	3	21
4–5	1	1	17	5	2	22	0	0	19	5	0	24
5–6	1	1	17	5	2	22	0	0	20	4	0	24
6–7	2	3	11	8	5	19	0	0	21	3	0	24
Total	9	6	54	27	15	81	3	0	75	18	3	93

(continued)

a) How many
   i) size classes were there
   ii) trials were made in total
   iii) limpets took part in the experiment?
b) After 30 minutes what can you conclude about the surface the limpets preferred?
c) What has happened in the crystallising dish during the period from 30 to 60 minutes?
d) What can you conclude about the limpets' surface preferences?
e) Do different sized limpets have different surface preferences?
f) A limpet has a foot like a snail or slug. Why do you think it prefers the surfaces shown in the results?
g) If the experiment was extended to 2 hours what results would you predict?

3 The tufted hair grass forms a clump called a tussock. This provides a habitat for different kinds of invertebrates. The numbers of individuals of the different kinds of invertebrates were investigated over an 8-month period.

Fifteen tussocks were dug up at random on a common each month and were taken apart carefully. The animals in each tussock were collected, arranged into groups and counted. The graphs in Figure 8.22, (opposite) were produced from the data collected.
a) How did the numbers of butterflies and moths (larvae and pupae) change during the 8 months?
b) Describe the population of animals in the tussocks in April.
c) How did the population change by May?
d) Why were the tussocks picked at random?

The animals found in the tussocks of tufted hair grass were compared with those in the tussocks of cocksfoot grass in May. The results were displayed in a bar chart (Figure 8.21).

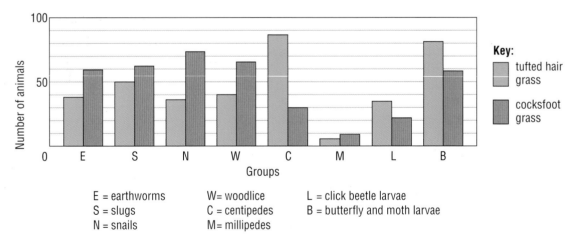

Figure 8.21

e) What are the two most numerous groups of animals in
   i) the tufted hair grass tussocks
   ii) the cocksfoot tussocks?
f) Which group of animals is found almost in the same numbers in both tussocks?
g) Which group of animals is twice as numerous in one type of tussock compared to the other?

(continued)

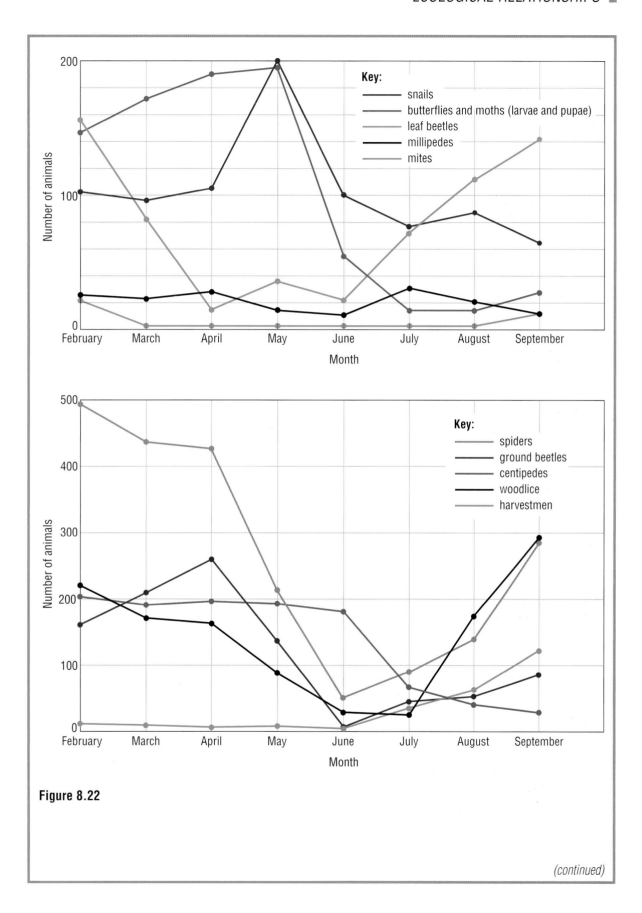

Figure 8.22

(continued)

The animals found in the tussocks of the cocksfoot grass in June were compared with those found in a rush tussock. The two pie charts in Figure 8.23 show how the populations compare.

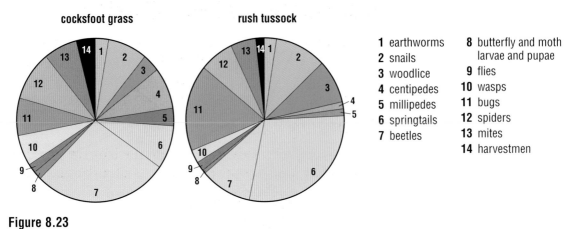

cocksfoot grass   **rush tussock**

1 earthworms
2 snails
3 woodlice
4 centipedes
5 millipedes
6 springtails
7 beetles

8 butterfly and moth larvae and pupae
9 flies
10 wasps
11 bugs
12 spiders
13 mites
14 harvestmen

Figure 8.23

h) What are the major ways in which the animal populations in the two tussocks differ?

9 Inheritance and selection

A closer look at cell division

Living things are made up of cells. It is the nucleus in a cell that controls the growth and development of the cell. The nucleus also provides the cell with all the instructions to carry out life processes.

When body cells divide

When a cell in a body divides, the nucleus divides first. Each new nucleus forms the nucleus of a daughter cell.

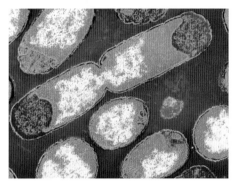

Figure 9.1 Two cells almost divided.

Just before a cell divides, long strands of material appear in the nucleus. These strands are called chromosomes.

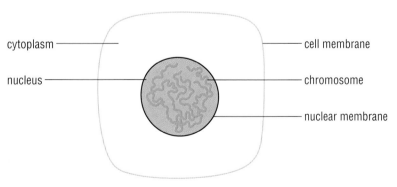

Figure 9.2 Chromosomes in the nucleus of a cell.

1 Draw a sequence of pictures about a cell which has only one chromosome from the time before the chromosome appears in the cell until two new daughter cells are produced and the chromosomes have disappeared into their nuclear material.

As the nucleus begins to divide, each chromosome makes a copy of itself. They lie together as two threads. During division of the nucleus each thread separates from the other and enters one of the new nuclei. Here it forms a chromosome. This means that the nuclei of the daughter cells have exactly the same number of chromosomes as the nucleus of the parent cell.

The chromosomes in the new nuclei of the daughter cells disappear back into the nuclear materials but reappear again when the cells are about to divide.

When gametes form

Gametes, such as eggs and sperm, are the sex cells of an organism. They join together in a process called fertilisation to produce a zygote (see page 40). If the gametes were produced in the same way as ordinary body cells, they would have the same number of chromosomes as body cells. This would create a problem at fertilisation, as the zygote would have twice as many chromosomes as its parents. If zygotes from this generation formed individuals which bred, the zygotes produced would have four times the number of chromosomes as the grandparents. Within a few generations the zygotes produced would be so packed with chromosomes that they would die. This situation does not develop in real life because the gametes are produced by a different kind of cell division.

When gametes are produced in the reproductive organs the cell division follows this sequence. In the body cell that is to produce gametes the chromosomes appear. In every body cell there are pairs of chromosomes. Each pair has a similar appearance to the other (the sex chromosomes are the exception, see page 166) and in this type of cell division the pairs come to lie side by side.

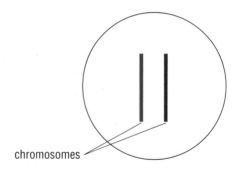

chromosomes

Figure 9.3 Two chromosomes in a nucleus.

Each chromosome then makes a copy of itself. We can think of the chromosome and its copy as two threads.

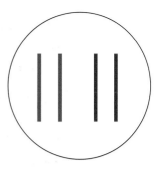

Figure 9.4 Two threads of two chromosomes.

A thread from one chromosome then swaps portions with a thread from the other chromosome.

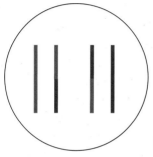

Figure 9.5 Threads after swapping material.

After this exchange of parts, the threads move apart. The cell divides to produce four daughter cells, which are the gametes. The nucleus of each gamete has a thread from one of the pair of chromosomes.

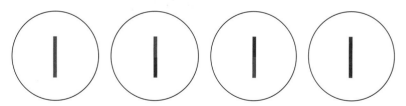

Figure 9.6 Gametes.

Each thread becomes a chromosome. This means that each gamete has only one chromosome, where the body cell had a pair. The division has halved the number of chromosomes.

When fertilisation occurs the nuclei of the gametes, each containing half the number of chromosomes of the normal body cells, join together. The nucleus of the zygote that is formed has the same number of chromosomes as the body cells of the parents. This means that the number of chromosomes in a zygote does not increase with each generation.

Each species has a certain number of chromosomes in the body cells. Human body cells have 46 chromosomes. Figure 9.7 shows how chromosomes pair up at fertilisation, and how an embryo develops with body cells that have the same number of chromosomes as the parent.

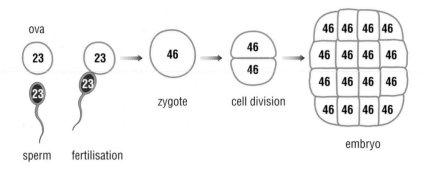

Figure 9.7 How chromosomes pair up at fertilisation.

All four gametes formed by a sperm-producing cell develop into sperm, but only one gamete formed from an ovum-producing cell develops into an egg or ovum.

A closer look at variation

In Chapter 4 we looked at how features vary within a species. Here we will look at how the features relate from one generation to the next. Figure 9.8 shows the faces of the parents and four children in a family. Features seen in the parents are also seen in the children, but the children are not exact copies of their parents.

Figure 9.8 Variation in a family.

2 What features have
    a) Ashley,
    b) Bryony,
    c) Charles,
    d) Davina inherited from their mother and father?

The features, such as hair colour and shape (wavy or straight), eye colour and the presence or absence of freckles and ear lobes, are present in both generations and have been passed from one generation to the next.

Chromosomes and genes

Before we look at how features are inherited we need to look at how the features are produced. The answer lies in the structure of the chromosomes. Chromosomes are really threads of chemical messages. The messages are strung along the chromosome like carriages in an intercity train. Each message is called a gene. The genes provide all the information for how a cell grows, develops and behaves.

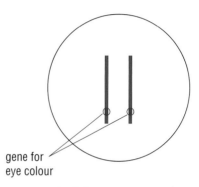

gene for eye colour

Figure 9.9 Two genes on a pair of chromosomes.

It is easy to imagine the genes of a single-celled organism such as *Amoeba* carrying out their tasks to keep the one cell alive, but what happens in a multi-cellular organism? Think of a human zygote. Its chromosomes contain all the genes to make the human body and keep it alive. How do these genes work together? The answer is that as the cells divide and form the fetus, all the genes in their nuclei are not used or switched on at once. When it is time for an organ to develop, the genes that control its development switch on and the cells that are produced make the organ. Other genes also present on the chromosomes but controlling the development of other organs do not switch on in these cells. For example, in the production of the windpipe the genes of some cells will switch on and make the cells into ciliated epithelium cells (see page 11). Their nuclei still contain other genes, such as for eye colour, but they remain switched off forever. When the iris of the eye is forming, genes for eye colour in the iris-making cells switch on while the genes for making cilia (microscopic hairs) remain permanently switched off.

How genes pass on

When gametes are made the genes pass into them on the chromosomes. During the formation of the gametes parts of the chromosomes swap portions. This swapping leads to a mixing up of the genes, so an exact copy of the parent's genetic code is not passed on.

When a zygote is formed at fertilisation, the nucleus contains all the genes needed to make a new individual. As there has been some mixing of the genes from both parents, the new individual develops a slightly different combination of features from their parents.

Mendelian genetics

Gregor Mendel (1822–1884) was an Austrian monk who studied mathematics and natural history. He set up experiments to investigate how features in one generation of pea plants were passed on to the next. Pea flowers self-pollinate. When Mendel wished to control the way the flowers pollinated he cut off the anthers of one flower, collected pollen from another flower and brushed it on to the stigma of the first. He completed his task by tying a muslin bag around the first flower to prevent any other pollen from reaching it.

Mendel performed thousands of experiments and used his mathematical knowledge to set out his results and to look for patterns in the way that the plant features were inherited. He suggested that each feature was controlled by an inherited factor. He also suggested that each factor had two sets of instructions and that parents pass on one set of instructions each to their offspring. Many years later it was discovered that Mendel's 'factors' were genes.

Mendel's work was published by a natural history society but its importance was not realised until 16 years after his death. At that time Hugo de Vries (1848–1935), a Dutch botanist, had been studying how plants pass on their characteristics from generation to generation. He was checking through published reports of experiments when he discovered Mendel's work. De Vries's work supported Mendel's but he also made another discovery. He had studied the evening primrose, a plant that had been newly introduced into Holland, and found that the plants occasionally produced a new variety that was quite different from the others. This new variety was caused by mutation. Although mutations had been seen in herds of livestock before, where the odd animal was sometimes called a 'sport', de Vries was the first to introduce the idea of mutations into scientific studies.

1 Why did Mendel cut out the anthers of some flowers?
2 Why did Mendel tie a muslin bag around the flower in the experiment?
3 What is the value of performing a large number of experiments?
4 How did Mendel's conclusions about factors compare with the discovery of how chromosomes behave when gametes are made?
5 Could money be saved on fencing if you farmed short-legged sheep? Explain your answer.
6 Mutation means change. Why is it a good word to describe a new variety of a species?

Figure A   Short-legged sheep mutation.

DNA

Genes are made from a substance called deoxyribonucleic acid which is usually shortened to DNA. The first work on investigating the chemicals in cell nuclei was carried out in 1869 by Johann Friedrich Miescher (1844–1895). He used the white cells in pus and the substance he discovered was called nuclein. Over the next 84 years generations of scientists made investigations on this substance. Rosalind Franklin (1920–1958) studied the structure of molecules by firing X-rays at them (see also Dorothy Hodgkin, page 132). In 1951 she investigated DNA in this way and her results suggested to her that it could be made of two coiled strands, but she was not sure. In 1953 James Watson and Francis Crick, using some of Franklin's results to help them, worked out that DNA is made from long strands of chemicals that are coiled together to make a structure called a double helix. The chemicals are arranged in a sequence that acts as a code. The code provides the cell with instructions on how to make the other chemicals that it needs to stay alive and develop properly.

Barbara McClintock (1902–1992) was a geneticist who studied maize – the plant that provides sweetcorn. While she was still a student she worked out a way of relating the different chromosomes in the nucleus to the features of the plant. Later, in the 1940s, she discovered that the genes on a chromosome could change position. They became known as 'jumping genes'. This discovery did not fit in with the way genes were thought to act and her work was not accepted by other scientists. But in the 1970s, during investigations by scientists on the DNA molecule, it was found that parts of the DNA broke off and moved to other parts of the chromosome. McClintock's work was proved to be correct and in 1983 she received the Nobel Prize for Physiology and Medicine.

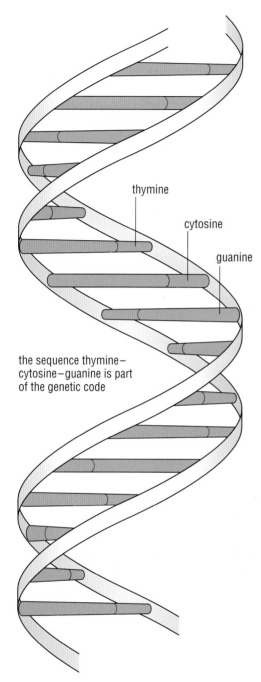

thymine

cytosine

guanine

the sequence thymine–cytosine–guanine is part of the genetic code

Figure A Basic structure of DNA.

Figure B Barbara McClintock.

(continued)

As each person's DNA is unique it can be used for identification purposes. A person's DNA profile (sometimes called a DNA fingerprint) can be made from cells in the saliva or the blood. The DNA is chopped up by enzymes and its pieces are separated into a gel in a process like chromatography. (Remember that chromatography is the process used to separate colours in an ink by putting a drop of ink onto a paper and allowing water to soak through it.) The pattern of the pieces looks like a bar code on an item of goods. Closely related people have more similar profiles than those who are not related.

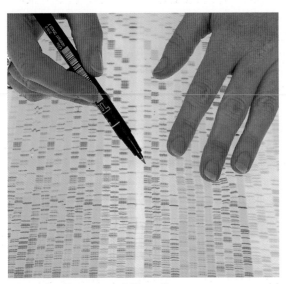

Figure C Examining a DNA profile.

For discussion

How could DNA be used to investigate a crime?

How firmly should scientists hold their views?

How males and females are produced

One pair of chromosomes are the sex chromosomes. There are two sex chromosomes and in humans they have different lengths. The longer one is called the X chromosome and the shorter one is called the Y chromosome. Females have two X chromosomes and males have an X and a Y chromosome. When the male makes sperm cells each one receives either an X or a Y chromosome as the pair divides. All the eggs receive an X chromosome because female cells do not contain a Y chromosome.

When an egg meets a sperm at fertilisation it has an equal chance of meeting a sperm containing an X or a Y chromosome because the sperms with the different sex chromosomes are produced in equal amounts.

3 What will be the sex of a baby produced when a sperm containing a Y chromosome fertilises an egg? Explain your answer.

4 Why do people not have two Y chromosomes?

5 Are girls or boys more likely to be formed after fertilisation? Explain your answer.

Species and varieties

In Chapter 4 we saw that living organisms are divided into groups called species. All the members of a species have the same number of chromosomes in their body cells. For example dogs have 78 chromosomes while cats have only 38; potatoes have 48 while peas have only 14.

When the males and females in a species reproduce they produce offspring which have the same number of chromosomes as themselves. The offspring are also capable of reproducing.

Most species have means which prevent them from trying to breed with other species. For example, if the pollen from one plant lands on the stigma of a different species it may not produce pollen tubes (see pages 20 and 22) because the correct concentration of sugar is not present. The behaviour of cats and dogs prevents them from breeding, but even if their sperm and eggs were mixed the difference in the number of their chromosomes would prevent them from combining and producing offspring with the same number of chromosomes as their parents.

However, within a species different varieties can be produced. The members of different varieties of a species have the same number of chromosomes as each other but they have different combinations of genes which give a different combination of features. For example, by carefully breeding tomato plants, tomatoes can be produced in red, orange and yellow. They can also be produced as small as cherries or as large as grapefruit to suit the requirements of different people. This is an example of selective breeding.

Figure 9.10 Different varieties of tomato.

Selective breeding

For thousands of years people have been breeding animals and plants for special purposes. Most plants were originally bred to produce more food but later plants were also bred for decoration. Animals were originally bred for domestication, then for food production or to pull carts.

A breeding programme involves selecting organisms with the desired features and breeding them together. The variation in the offspring is examined and those with the desired feature are selected for further breeding. For example, the 'wild' form of wheat makes few grains at the top of its stalk. Individuals that produce the most grains are selected for breeding together. When their offspring are produced they are examined and the highest grain producers are selected and bred together. By following this programme wheat plants producing large numbers of grains have been developed.

Figure 9.11 'Wild' wheat (left) and modern wheat (right).

In some breeding programmes a number of features are selected and brought together. The large number of different breeds of dog have been developed in this way.

6 All the different breeds of dog have been developed from the wolf by selective breeding. What features do you think have been selected to produce a greyhound? Give a reason for each feature you mention.

Figure 9.12 A wolf and a greyhound.

Natural selection

For discussion

How do you think that the human species may evolve in the future?

Just as humans select individuals for further breeding, it has been found that selection in a habitat takes place naturally. Those individuals of a species which have the most suitable features to survive will continue to live and breed and pass that their features. Individuals which have features that do not equip them for survival will perish, and so do not pass on their features to future generations. This form of selection is used to explain the theory of evolution, where one species changes in time until another species is produced. The finches on the Galapagos Islands, first studied by Charles Darwin (1809–1882), are thought to have evolved by natural selection (see Figure 9.13 overleaf).

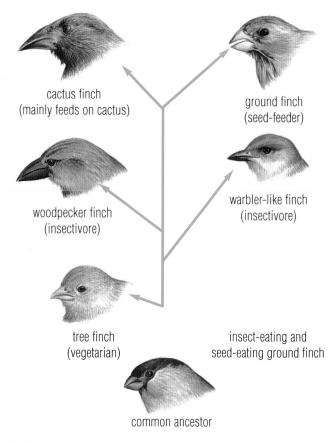

cactus finch
(mainly feeds on cactus)

ground finch
(seed-feeder)

woodpecker finch
(insectivore)

warbler-like finch
(insectivore)

tree finch
(vegetarian)

insect-eating and
seed-eating ground finch

common ancestor

Figure 9.13 Darwin's finches

Clones

A clone is an organism that is an exact copy of its parent. Clones can occur naturally through a process called asexual reproduction. *Amoeba* reproduces asexually when it simply divides in two. *Hydra*, a small animal related to sea anemones which lives in ponds and ditches, grows a bud which detaches itself and becomes a copy of its parent. Some plants can also form clones. The spider plant is a familiar houseplant which grows small plants on side shoots (see page 69). The small plants can become detached and live on their own. A simple way of cloning a plant is to take a cutting from it and grow the cutting in compost to make a new plant.

A technique for cloning animals has been developed using nuclei and cells. The nucleus is removed from an egg cell and is replaced by the nucleus of a normal body cell from the animal you wish to clone. The egg is then allowed to develop normally and an exact copy of the animal is produced. Dolly the sheep was the first successfully reared clone at the end of the 20th Century. Since then many other clones have been made.

For discussion

How could cloning help farming?

What disadvantages might there be to cloning farm animals?

If you were cloned today would your clone be just like you when it is your age? If not, why not?

◆ SUMMARY ◆

◆ The nucleus of a cell contains chromosomes (*see page 159*).

◆ When a cell divides each new nucleus receives chromosomes from the parent cell (*see page 159*).

◆ A special type of cell division occurs in the reproductive organs that makes sure the following generation has the same number of chromosomes in their cells as the parent generation (*see page 160*).

◆ There are genes on chromosomes which carry instructions for the cell (*see page 163*).

◆ The sex of an individual is determined by the sex chromosomes (*see page 166*).

◆ There are varieties within species (*see pages 166–167*).

◆ New varieties of a species can be produced by selective breeding (*see page 168*).

◆ Natural selection is the process by which new species evolve (*see page 169*).

End of chapter questions

1 a) The body cell of a wheat plant has 42 chromosomes. How many chromosomes are in the nucleus of an ovule?

b) The body cell of a chicken has 78 chromosomes. How many chromosomes are in the nucleus of a sperm cell?

c) The nucleus of a pollen grain of a cabbage has 9 chromosomes. How many chromosomes are in the nucleus of a body cell?

d) The egg of a housefly has 6 chromosomes. How many chromosomes are in the nucleus of a body cell?

2 Camouflage helps animals survive in their habitat through the process of natural selection. Investigate camouflage by studying the results of this experiment.

Garden birds were given different coloured food pellets against different coloured backgrounds to test the effect of camouflage. The food pellets were made of flour and lard and were coloured with harmless dyes. The pellets represented prey animals and were coloured either green or brown. The prey were tested against three backgrounds – grey, green and brown. The backgrounds were painted on 50 cm square aluminium sheets.

Twenty-four trials were made, twelve with grey backgrounds and six each with green and brown backgrounds. In each trial two backgrounds of the same colour were set up a metre apart in a garden. Ten green and ten brown prey were placed on each sheet. When half the prey had been eaten by the garden birds the trial was stopped and the numbers of green and brown prey that had been eaten were recorded. The trials with pairs of coloured backgrounds were made in a random order. The results of the trials are displayed in Table 9.1.

(continued)

Table 9.1

Grey background				Green background		Brown background	
Numbers eaten		Numbers eaten		Numbers eaten		Numbers eaten	
Green	Brown	Green	Brown	Green	Brown	Green	Brown
18	10	8	19	6	17	11	2
11	8	14	5	0	11	10	2
8	13	7	11	15	4	19	1
12	4	15	6	12	12	5	7
18	9	14	4	12	19	12	8
4	12	18	17	11	9	5	11
Grand totals:		147	118	56	72	62	31

a) Why were two backgrounds used in each trial?

b) Why were the backgrounds mixed up at random and not put out as grey twelve times, green six times and brown six times?

c) Were green or brown prey more likely to be eaten on the grey background?

d) Identify any pairs of numbers that do not follow this trend.

e) What percentage of green prey are eaten on a grey background?

f) What is the percentage of green prey eaten on
   i) a green background and
   ii) a brown background?

g) How might you have expected the results on the green background to be different from those in the table?

h) Do the colours of the prey help to camouflage them? Explain your answer.

10 Keeping healthy

A body is made up of billions of cells. Groups of similar cells form tissues that work together with other tissues in larger structures called organs. The organs form groups called organ systems (see page 1) which work together to keep the body alive. The power to make the organ systems work comes from the energy in food, which is released by chemical reactions. The way the body is made and the way it works are incredibly complicated. If you live a healthy lifestyle and have a healthy diet (see Chapter 5) your body may keep working for over 80 years. If you live an unhealthy lifestyle the cells, tissues and organs may become damaged.

Figure 10.1   A family enjoying a balanced meal.

Exercise

Regular exercise makes many of the organ systems become more efficient. It also uses up energy and helps to prevent large amounts of fat building up in the body. Exercise can increase your fitness in three ways: it can improve your strength, make your body more flexible and less likely to suffer from sprains, and it can increase your endurance which is your ability to exercise steadily for long periods without resting. Different activities require different levels of fitness. Table 10.1 shows these levels for different sporting activities. By studying the table you can work out which activities you could do to develop one or more of the three components of fitness.

1 Which activities demand great flexibility?
2 Which activity is the least demanding?
3 Which activities are the most demanding?
4 How do the demands of soccer and long distance running compare?
5 Which activity would you choose from the table? What are its strengths and weaknesses?
6 Many people claim that they do not have time to exercise. How would you motivate such people to take some form of exercise? Which activities might suit them best?

Table 10.1

Activity	Strength	Flexibility	Endurance
Basketball	✔✔	✔✔	✔✔✔
Dancing	✔✔	✔✔✔	✔✔
Golf	✔✔	✔✔	✔✔
Long distance running	✔✔✔	✔✔	✔✔✔
Soccer	✔✔	✔✔	✔✔✔
Squash	✔✔✔	✔✔✔	✔✔✔
Swimming	✔✔✔	✔✔	✔✔✔
Tennis	✔✔✔	✔✔✔	✔✔✔
Walking	✔	✔	✔✔

Smoking and health

In Chapter 6 we saw how the respiratory system works to provide us with an exchange of respiratory gases. An efficient exchange is needed for good health. When people smoke they damage their respiratory system and risk seriously damaging their health.

There are over a thousand different chemicals in cigarette smoke, including the highly addictive nicotine. These chemicals swirl around the air passages when a smoker inhales and touch the air passage linings. In a healthy person, dust particles are trapped in mucus and moved up to the throat by the beating of microscopic hairs called cilia. The small amounts of dust and mucus are then swallowed. In a smoker's respiratory system the cilia stop beating due to chemical damage by the smoke. More mucus is produced but instead of being carried up by the cilia it is coughed up by a jet of air as the lungs exhale strongly. This is a smoker's cough and the amount of dirty mucus reaching the throat may be too much to swallow.

In time chronic bronchitis may develop. The lining of the bronchi become inflamed and open to infection from microorganisms. The inflammation of the air passages makes breathing more difficult and the smoker develops a permanent cough. The coughing causes the walls of some of the alveoli in the lungs to burst. When this happens the surface area of the lungs in contact with the air is reduced. This leads to a disease called emphysema.

Some of the cells lining the air passages are killed by the chemicals in the smoke. They are replaced by cells below them as they divide and grow. Some of these cells may be damaged by the smoke too and as they divide they may form cancer cells. These cells replace the normal cells in the tissues around them but they do not perform the functions of the cells they replace. The cancer cells continue to divide and form a lump called a tumour. This may block the airway or break up and spread to other parts of the lung where they can set up more tumours.

7 What is the function of a smoker's cough?

8 Why may chronic bronchitis lead to other diseases?

9 How does the reduced number of alveoli affect the exchange of oxygen and carbon dioxide?

10 Why does someone with emphysema breathe more rapidly than a healthy person?

11 How are cancer cells different from normal cells in the lung tissue?

12 Why do cancer cells in an organ make the organ less efficient?

13 Why might the growth of cancer tumours in an organ have fatal results?

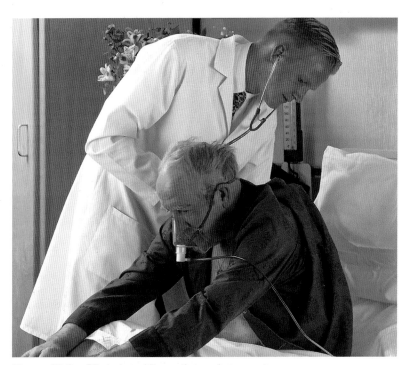

Figure 10.2   Effect of smoking on the respiratory system.

A healthy diet

In Chapter 5 we looked in detail at the importance of diet. From this work you can think about the amounts of food you eat in the following way.

You can eat large amounts of potatoes, bread, rice and pasta. They provide you with carbohydrates which supply the body with energy. You should eat a smaller amount of fruit and vegetables (but still five portions a day) to provide you with vitamins, minerals and fibre. You should eat a smaller amount still of foods such as meat and fish, to provide you with the protein you need for growth and repair of the body. In a vegetarian diet protein is mainly provided by pulses such as beans, lentils and peas. Finally you should eat even smaller amounts of food rich in fat such as chocolate, nuts, fatty meat and cheese. Fat provides materials for making cell membranes. It also creates an insulating layer beneath the skin that helps to retain body heat. This layer acts as an energy store for the body, but there can be dangers to health if it becomes too thick.

An easy way to think about all this information is to think of a pyramid of food, as Figure 10.3 shows.

Figure 10.3   A pyramid of food.

14 Make a pyramid of food representing your diet. Describe how it compares with the pyramid in Figure 10.3. If it does not match the pyramid in the diagram, how can you change your diet to make it healthier?

Note: this pyramid of food is simply to make you think about eating healthily. It is not related to ecological pyramids found on page 146.

The pyramid reminds you that you should eat large amounts of foods at the bottom of the pyramid but only small amounts of those at the top.

Alcohol

Many people have their first alcoholic drink at special family occasions such as birthdays and weddings. As they get older they may have a few drinks at weekends then perhaps during the week too. Drinking alcohol may be used at first as a form of relaxation with friends but may become a regular daily habit until the person cannot cope without drinking large amounts daily. This final stage is alcoholism. To avoid this people need to develop a sensible attitude towards alcohol.

Effect of alcoholic drinks

Alcohol affects the nervous system. It slows down the speed at which nerve cells carry signals. A small amount of alcohol may make a person feel more relaxed but it also makes the nerves work slightly more slowly. This makes a person react more slowly. As the person drinks more, the effect of the alcohol on the nervous system increases and their reaction time becomes longer. The behaviour of the person may change. Their voice may become louder and they may become reckless and even aggressive. The person finds it more difficult to think and speak clearly. If they continue to drink their body movements may become unco-ordinated and the person may be unable to walk. They may fall asleep or become unconscious. In extreme cases, in the unconscious state they may be sick and if the vomit gets stuck in their windpipe they may suffocate and die.

Long-term effects of alcohol abuse

Alcohol is a poison. The liver collects poisons from the blood as it flows through. It breaks down the poisons to make them harmless. If large amounts of alcohol are drunk over many years the liver may become inflamed and develop a disease called hepatitis. Parts of the liver may turn to scar tissue. This leads to the development of cirrhosis of the liver which reduces the liver's capacity to neutralise poisons. This disorder can be fatal.

Strength of alcoholic drinks

Bottles and cans of alcoholic drinks have the strength of their alcohol content marked on them. It is shown as a number with % ABV written after it. This means % of alcohol by volume. Many beers have a strength of 3.5% ABV, while cans of lager or cider may be much stronger, having a strength of 9% ABV. Whisky may be 40% ABV, while sherry is about 15% ABV and wine about 11% ABV. Some 'alcopops' may be up to 13.5% ABV.

Measures of alcohol

Alcohol is measured in units of pure alcohol. One unit is 8 g or 10 cm³ of pure alcohol. To help people assess how much alcohol they are drinking the following three measures of drinks are used.

A half pint (284 ml) of 3.5% ABV beer = 1 unit

A 25 ml measure of a spirit drink such as whisky at 40% ABV = 1 unit

A 125 ml glass of wine at 8% ABV = 1 unit (most wines have a higher % ABV than 8%)

The number of units in a drink can be calculated by using this formula:

$$\frac{\text{Volume of drink}}{1000} \times \% \text{ ABV} = \text{number of units}$$

For discussion

Would you be happy being driven by somebody who had recently drunk up to 5 units of alcohol?

15 Arrange wine, beer, 'alcopops', whisky, sherry and cider in order of their alcoholic content, putting the strongest one first.

16 Do you think 'alcopops' are suitable alcoholic drinks for young people? Explain your answer.

17 Calculate the units in these drinks:
   a) 440 ml can of lager at 4% ABV
   b) 500 ml can of cider at 9% ABV
   c) 20 cl bottle of 'alcopop' at 13.5% ABV.

18 A man has two pints of beer in an hour and his wife has three glasses of wine. How long does it take their livers to destroy the alcohol they have drunk?

19 How long would it take a liver to destroy all the alcohol in a 75 cl 10% ABV bottle of wine? Note: 1 cl = 10 ml.

Figure 10.4   Examples of drink units.

Using the measures to stay healthy

If people wish to drink and remain healthy they should not drink beyond a certain number of units of alcohol each week. For men the recommended limit is 21 units and for women the recommended limit is 14 units. The difference in limits is due to the size and water content of the body. Men are generally larger than women and have a higher water content. Between 55 and 65% of the body weight of a man is due to water, while water forms only between 45 and 55% of a woman's body weight.

Many countries have laws that aim to prevent people driving under the influence of alcohol. In the United Kingdom this limit is set at the amount of alcohol that gives a blood alcohol level of 80 mg/100 ml. The number of units that gives this blood alcohol level varies between different people but is about 3 units. In other countries the limit is less. Many people believe that no alcohol at all should be drunk before driving.

The liver and removal of alcohol

It takes the liver about 1 hour to destroy 1 unit of alcohol. If a person drinks more than 1 unit in an hour, the concentration of alcohol in the blood increases and affects the rest of the body.

Solvents

When they are sniffed, the chemicals in a range of solvents can produce an effect similar to drunkenness. They cause the sniffer to have weird sensations that may be frightening or make them behave in a foolish or dangerous way.

The solvents that are sniffed are usually in certain kinds of glues, correcting fluids and aerosol sprays. The butane gas used as the fuel in cigarette lighters is also inhaled. The effect of a solvent on the body occurs very quickly because the chemicals enter the blood through the thin lining of the lungs. The effect does not last long. If a person wants to stay under the effect of the solvent he/she has to keep sniffing.

People react differently to the chemicals in solvents. Some young people have died after sniffing solvents for the first time. The solvents can damage lungs, the control of breathing, the nervous system, liver, kidneys and bone marrow. As sniffers lose control of themselves, they may suffocate on the plastic bags they use to inhale the solvent, become unconscious then be sick and suffocate in their vomit, or cause a fire and risk burning themselves when lighting a cigarette near solvents that are flammable.

A healthy heart

The human heart starts to form in the embryo 20 days after conception (see page 44). Fewer than 1% of babies are born with heart defects, yet in the United Kingdom more people die from heart disease than from any other disease.

The heart may beat up to 2500 million times during a person's life. Its function is to push blood around the 100 000 km of blood vessels in the body. This push creates a blood pressure that drives the blood through the blood vessels. As the ventricles in the heart fill with blood the pressure in the blood vessels is reduced, but as the heart pumps it out along the arteries the blood pressure rises. The walls of the arteries are elastic and they stretch and contract with the blood pressure. In young people the artery walls are clear and their diameters are large enough to let the blood flow with ease. As the body ages the artery walls become less elastic.

The heart has its own blood vessels called the coronary arteries and veins. They transport blood to and from the heart muscle.

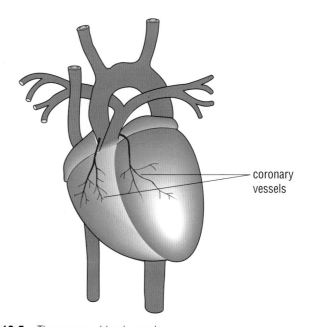

coronary
vessels

Figure 10.5 The coronary blood vessels.

Fatty substances, such as cholesterol, stick to the walls of arteries. Calcium settles in the fatty layer and forms a raised patch called an atheroma. The blood then has less space to pass along the arteries and its pressure rises as it pushes through the narrower tubes. Other components of the blood, such as platelets, settle on the atheroma and make it larger. This may cause a blood clot which narrows the artery even more or can completely block it causing a thrombosis. This means that the artery is unable to supply oxygen and other nutrients to the relevant organ. A thrombosis in a coronary artery causes a heart attack. A thrombosis in an artery in the brain causes a stroke.

The features that develop in the body that cause heart disease can be inherited. People whose relatives have suffered from heart disease should take special care to keep their heart and circulatory system healthy.

Keeping the heart healthy

The heart is made of muscle and like all muscles it needs exercise if it is to remain strong. The heart muscles are exercised when you take part in the activities in Table 10.1 (see page 174). Heart muscle contracts more quickly and more powerfully during exercise than it does at rest so that more blood can be pumped to your muscles. These muscles need more blood to provide extra oxygen while they work.

As we have seen, the blood supply to heart muscles can be reduced by fatty substances such as cholesterol in the blood. These substances are formed after the digestion of fatty foods. Some fatty substances are needed to keep the membranes of the cells healthy, but too much intake of fat leads to heart disease. A heart can be kept healthy by cutting down on the amount of fat in the diet. This may be achieved by cutting fat off meat, or eating fewer crisps and chips for example.

Care with exercise

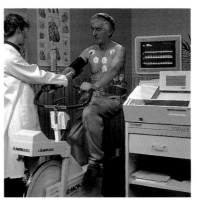

Figure 10.6 Pedalling an exercise bicycle makes the heart beat faster to provide blood for the leg muscles.

When people decide to get fit, they may choose one of the activities from Table 10.1 and begin with great enthusiasm. However, they may experience a sprain or pains by trying to exercise too hard too early. The skeleton and muscles work together to provide movement. Someone who has not been active for a long time may need to build up their exercises gradually so that the muscles and joints can become adapted to the increased activity. If this is not done and someone receives an injury early on in their exercise programme they may decide not to continue, and they will become unfit again. It can help to be aware of how the skeleton and muscles work, and to think of them when an exercise programme is begun.

20 There are three bones in the arm. How many are in the wrist and hand?

21 How many bones are in all four limbs?

22 A person has a mass of 43.5 kg. What is the mass of their skeleton?

23 Figure 10.7 shows the main bones of the body. How many of these bones can you feel in your body?

Skeleton

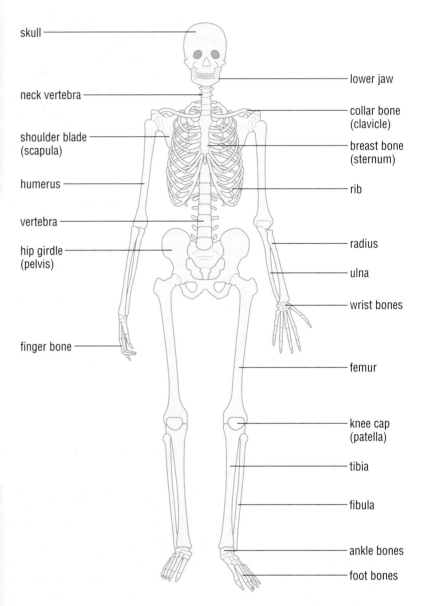

Figure 10.7 The skeleton.

24 The skull forms a solid sheet of protection and the ribs form a cage. Why do you think the rib cage is not a solid sheet like the skull? Which offers the better protection, the sheet or the cage? Explain your answer.

25 Newborn mammals have soft skeletons to allow some flexibility during birth. All mammals are fed milk by their mothers. What effect will this have on their bones?

There are 206 bones in the human skeleton. Each arm and hand together have 30 bones. Each leg and foot together have 29 bones. The skeleton accounts for 15% of the mass of the body. The tissue of the skeleton (bone) is hardened as it takes up calcium from the digested food.

The skeleton and protection

The brain and the spinal cord form the central nervous system and are made from soft tissue. They could be easily damaged without a hard covering. The bones of the skull are fused together to make a strong case around the brain. The backbone is made up of 33 bones known as vertebrae (*singular:* vertebra). There is a hole in each vertebra through which the spinal cord runs. The column of vertebrae makes a tube of bone around the spinal cord. There are gaps between the vertebrae through which nerves pass from the spinal cord to the body. The ribs and backbone form a protective structure around the lungs and the heart.

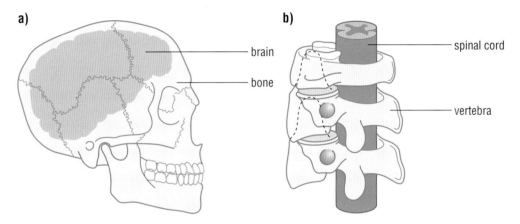

Figure 10.8 Protection of the central nervous system, **a)** the brain and **b)** the spinal cord.

26 A person has a mass of 43.5 kg. How much of this mass is due to
  a) their organ systems and
  b) their muscles?

27 The percentage of the body's mass not accounted for by the skeleton, organs and muscles is due to fat. What percentage of the body's mass is due to fat?

The skeleton and support

The organs that form systems such as digestive, circulatory, excretory and respiratory systems account for 20% of the body's weight. The organs are made from soft material and have no strong supporting material inside them. The bones of the skeleton provide a strong structure to which the organs are attached. They allow the organs to be spread out in the body without squashing into each other.

The muscles account for 45% of the body's weight. They are also made from soft tissue but gain their support from the bones to which they are attached.

28 Look at the skeleton in Figure 10.7 (page 182). Which bones meet at
  a) the hip joint,
  b) the knee joint,
  c) the elbow joint and
  d) the shoulder joint?
29 Name
  a) two hinge joints and
  b) two ball and socket joints.
30 How might a joint be affected by
  a) torn ligaments,
  b) lack of synovial fluid and
  c) damaged cartilage?

31 Why do you think that some joints are painful in elderly people?
32 How does the body stop you using a damaged joint so that it has time to heal?

For discussion
What would the body be like without a skeleton?

Could the body survive without a skeleton?

The skeleton and movement

The place where bones meet is called a joint. In some joints, such as those in the skull, the bones are fused together and cannot move. Most joints, however, allow some movement. Some joints such as the elbow or knee are called hinge joints, because the movement is like the hinge on a door. The bones can only move forwards or backwards. A few joints, such as the hip, are called ball and socket joints because the end of one bone forms a round structure like a ball which fits into a cup-shaped socket. These joints allow much more movement.

To stop the bones coming apart when they move, they are held together by tough fibres called ligaments. To stop them wearing out as they rub over each other, the parts of the bones in the joint are covered with cartilage. This substance has a hard, slippery surface that reduces friction and allows the bones to move over each other easily. In some joints, where there is a lot of movement, cells in a tissue called the synovial membrane make a liquid called synovial fluid. This fluid spreads out over the surfaces of the cartilage in the joint and acts like an oil, reducing friction and wear.

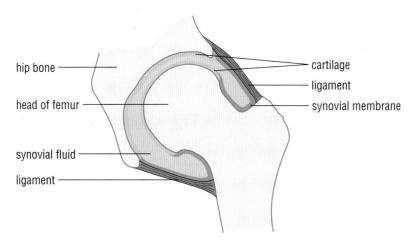

Figure 10.9 Inside a hip joint.

Muscles

Muscle is made up from tissue that has the power to move. It can contract to become shorter. A muscle is attached to two bones across a joint. When muscle gets shorter it exerts a pulling force. This moves one of the bones but the other stays stationary. For example, in the upper arm the biceps muscle is attached to the shoulder blade and to the radius bone in the forearm.

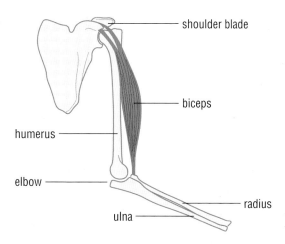

Figure 10.10 Biceps on arm bones.

33 Draw a diagram featuring both the biceps and the triceps, showing the triceps fully shortened.
34 Using dotted lines, draw the position of the forearm when the biceps is fully shortened.

When the biceps shortens or contracts it exerts a pulling force on the radius and raises the forearm.

A muscle cannot lengthen or extend itself. It needs a pulling force to stretch it again. This force is provided by another muscle. The two muscles are arranged so that when one contracts it pulls on the other muscle, which relaxes and lengthens. For example, in the upper arm the triceps muscle is attached to the shoulder blade, humerus and ulna. When it contracts, the biceps relaxes and the force exerted by the triceps lengthens the biceps and pulls the forearm down. When the biceps contracts again, the triceps relaxes and the force exerted by the biceps lengthens the triceps again and raises the forearm. The action of one muscle produces an opposite effect to the other muscle and causes movement in the opposite direction. The two muscles are therefore called an antagonistic muscle pair.

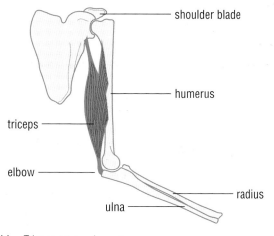

Figure 10.11 Triceps on arm bones.

Drug abuse

Drugs are chemical substances that change the way we think, feel or behave. Some, such as caffeine which is found in tea and coffee, are made by plants. Medical drugs are usually made by the pharmaceutical industry from raw chemical materials. Most drugs are produced to ease the symptoms of a disease or to cure the disease. Although many people think of harmful drugs as being substances like heroin or cannabis, nicotine in cigarettes and alcohol in drinks like beer and wine are also drugs that can harm the body.

People begin taking harmful drugs for a variety of reasons. Some begin because they are unhappy, lonely or feel that they are unable to cope with life. They think the drugs will make them feel better. Others take them because their friends are trying them and they find it difficult to say no (peer pressure). Other people take them because they think it is exciting to use substances that are illegal. Whatever the reason, taking harmful drugs can be dangerous.

General body changes in drug abuse

A drug has an effect on how the body works. As a person continues to take certain drugs the body becomes more tolerant of them and larger amounts of the drug have to be taken for the person to feel its effects. The drug-taker's brain or body generally gets so used to the drug that it becomes changed in some way and becomes physically dependent on the drug. This is known as an addiction. If the person stops taking the drug the body reacts in a range of painful ways, including sickness. These reactions are called withdrawal symptoms.

While the body is becoming physically dependent on the drug, the person may be becoming psychologically dependent on it. This means that they become upset if they are not taking it and develop the irrational fear that they cannot cope with life without the drug.

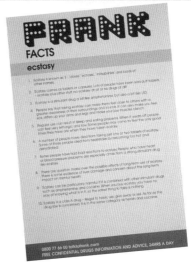

Figure 10.12

Dangers of drugs

Ecstasy

Ecstasy affects the body's co-ordination. It can make a person confused. It also acts as a stimulant and can seriously affect people who suffer from epilepsy or have a heart condition. It can be fatal.

Amphetamine or speed

Amphetamines are stimulants. Their use can lead to mental disorders and heart damage.

Cocaine and crack

Cocaine is sniffed or injected. Crack is a form of cocaine that can be smoked. Both are stimulants and are highly addictive. They can make users feel sick, itchy, suffer nose damage, have difficulty sleeping and develop mental disorders.

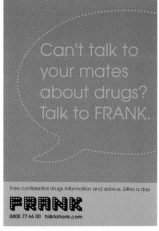
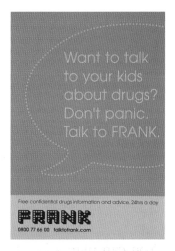
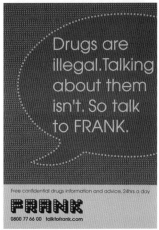
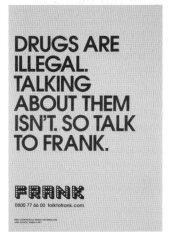

Figure 10.13 FRANK is a government campaign that offers advice on how drugs can affect you.

For discussion

What might be the effect of making the drugs in this section legal?

How effective are posters at preventing drug abuse?

What else could be useful?

LSD (acid)

LSD affects the brain and makes the user see things that are not there. These illusions are called hallucinations. They may be frightening and lead to the person being upset when the effect has worn off. People who use LSD regularly can become less alert to the world around them.

Cannabis

Cannabis can produce hallucinations and make people upset. If it is smoked it can cause the same diseases as smoking tobacco.

Heroin

Heroin can make first-time users so sick that they will never try it again. Regular heroin users are physically and psychologically dependent on it and will commit crimes to obtain money to buy more. They may have slurred speech and seem slightly sleepy. Heroin addicts who inject the drug are in danger of catching hepatitis and AIDS from sharing dirty needles. When heroin addicts try to give up the drug they have great difficulty losing their psychological dependence. They may still crave for the drug after they have stopped taking it for a long time.

◆ SUMMARY ◆

- Regular exercise helps the body to stay healthy (*see page 173*).
- Smoking damages the respiratory system (*see page 174*).
- You can plan a healthy diet if you think of the different foods arranged in a pyramid (*see page 176*).
- Alcohol abuse produces changes in behaviour which can be life-threatening. Prolonged abuse causes liver damage (*see page 177*).
- Solvents damage the lungs, nervous system, kidneys and bone marrow. They cause changes in behaviour that can be life-threatening (*see page 179*).
- Diet can affect the supply of blood to the heart (*see page 181*)
- Muscles and bones work together to provide movement (*see page 184*)
- Illegal drugs can cause damage to a wide range of body organs and can lead to mental disorders (*see pages 188–189*).

End of chapter questions

1 How is someone putting their health at risk when they smoke tobacco daily?

2 Make a table about the drugs featured on pages 188–189 under the headings

- Drugs
- Effects on the body
- Effects on behaviour

3 Design a poster based on the information in your table. Compare the posters produced in your class, and rank them for effectiveness of putting people off trying drugs.

4 Some people have to inject themselves once a day to stay healthy. Can you think of a condition where this is the case, and name the substance that is injected?

For discussion

What is an ideal lifestyle for a healthy life?

11

How green plants live

How experiments build up information

Scientific processes are not usually understood by a single activity or even one repeated several times for checking. Processes are worked out over many years by a large number of different experiments that require different apparatus and different techniques. They may include making observations, thinking up new ideas and making models to test ideas. The activities form part of a line of research that may go back many years. The results of each experiment may contribute something to our understanding of how a process works. Eventually the results of a large number of different activities may show how the process works. In the following pages a series of experiments are presented as very simple examples of how their results contributed to our understanding of how plants make food.

If you become a scientist you will use some of the features you read about here, such as looking at the work of others or learning a technique to use in your experiments, in addition to making investigations following the scientific method.

The willow tree experiment

In the 17th century Joannes Baptista van Helmont (1580–1644) performed an experiment on a willow tree. He was interested in what made it grow. At that time scientists believed that everything was made from four 'elements': air, water, fire and earth. Van Helmont believed that water was the most basic 'element' in the universe and that everything was made from it. He set up his experiment by weighing a willow sapling and the soil it was to grow in. Then he planted the sapling in the soil and provided it with nothing but water for the next five years. At the end of his experiment he found that the tree had increased in mass by 73 kilograms but the soil had decreased in mass by only about 60 grams. He concluded that the increase in mass was due to the water the plant had received.

1 How fair was van Helmont's experiment? Explain your answer.
2 Did the result of the experiment support van Helmont's beliefs? Explain your answer.
3 If you were to repeat van Helmont's experiment how would you improve it and what table would you construct for recording your results?

Figure 11.1 Watering a willow tree.

If we were to summarise his conclusion it could look like this:

$$water \rightarrow mass\ of\ plant$$

Revising the work so far

As plants are food for animals the simple equation could be rewritten as:

$$water \rightarrow food\ in\ the\ plant$$

Moving on

The idea of food in the plant could then be investigated. A reasonable place to start could be with a plant part that is used as food – the potato.

Examining a potato with a microscope

If a small slice of potato is examined under the microscope the cells are found to contain colourless grains. When dilute iodine solution is added to the potato slice the grains turn blue–black. This test shows that the grains are made of starch.

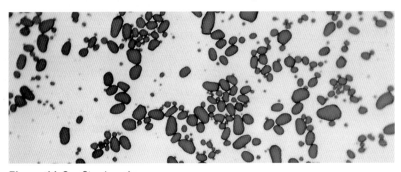

Figure 11.2 Starch grains.

Moving on

Having established that plant tubers such as the potato contain starch it may then be reasonable to try and find out if other parts of the plant contain starch. As the leaves are a major feature of most plants the search for starch in leaves would be the next task.

Testing a leaf for starch

Iodine does not produce a colour change when it is placed on a leaf because the cell walls will not allow the iodine into the cells and the green pigment masks any colour change. However, the work of others has shown that a leaf can be tested for starch if it is first treated with boiling water and ethanol. The boiling water makes it easier for liquids to leave and enter the plant cells and the ethanol removes the green pigment, chlorophyll, from the leaf and makes the leaf crisp.

If the leaves of a geranium that has been growing on a windowsill or in a greenhouse are tested, they will be found to contain starch.

4 What was the purpose of putting the leaf
a) in boiling water and
b) in ethanol?

Revising the work so far

Starch belongs to a nutrient group called carbohydrates, so, perhaps the simple equation could be altered to:

$$water \rightarrow carbohydrate\ (starch)\ in\ the\ plant$$

Moving on

Stephen Hales (1677–1761) discovered that 'a portion of air' helped a plant to survive and Jan Ingenhousz (1730–1799) showed that green plants take up carbon dioxide from the air when they are put in the light. It was also known that water contains only the elements hydrogen and oxygen while carbohydrates contain carbon, hydrogen and oxygen. All this information led to a review of van Helmont's idea that only water was needed to produce the carbohydrate. The review began by considering what else was around the plant apart from water. It was known from van Helmont's work that the soil contributed only a very small amount to the increased mass of the plant. The only other material coming into contact with the plant was the air. Ingenhousz's work suggested that the carbon dioxide in the air was important. This idea can be tested in the laboratory.

Destarching a plant

If you want to see whether starch has been made you have to start with a plant that does not have starch. If a plant that has leaves containing starch is left in darkness for two or three days, then tested again, it will be found that the leaves are starch-free. The plant is described as a destarched plant. It can be used to test for the effect of carbon dioxide.

Investigating the effect of carbon dioxide on starch production

Soda lime is a substance that absorbs carbon dioxide and takes it out of the air. Sodium hydrogencarbonate solution is a liquid that releases carbon dioxide into the air.

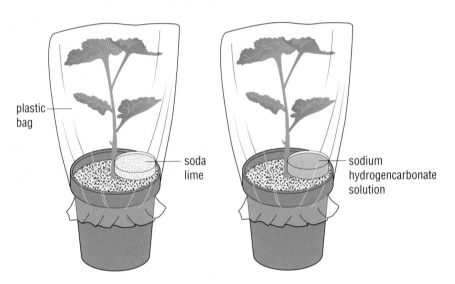

Figure 11.3 Plants set up to investigate the effect of carbon dioxide on starch production.

5 What does soda lime do to the air inside the plastic bag?

6 What does sodium hydrogencarbonate do to the air inside the plastic bag?

Two destarched plants were set up under transparent plastic bags that were sealed with an elastic band. Before covering the plants with the bags, a small dish of soda lime was added to one plant and a small dish of sodium hydrogencarbonate solution was added to the other. Both plants were left in daylight for a few hours before a leaf from each of them was tested for starch.

The leaf from the plant with the soda lime dish did not contain starch but the leaf from the plant with the sodium hydrogencarbonate did contain starch. This suggested that carbon dioxide is needed for starch production.

Revising the work so far

After reviewing the result of the effect of carbon dioxide on starch production the simple equation can be modified again to:

carbon dioxide + water → carbohydrate (starch) in a plant

Moving on

Joseph Priestley (1733–1804) studied how things burn. At that time scientists used the phlogiston theory to explain how things burned. They believed that when materials such as wood burned they lost a substance called phlogiston. When a candle burned in a closed volume of air, such as the air in a bell jar, they believed that the candle eventually went out because the air had become filled with phlogiston. It had become phlogisticated.

When Priestley put a plant in the air in which a candle had burned he found that later on a candle would burn in it again. He reasoned that the plant had taken the phlogiston out of the air and had made dephlogisticated air. Later, Ingenhousz re-examined Priestley's results and found that the phlogiston theory was wrong. The plants had in fact produced oxygen.

Water plants can be used to investigate the gases produced by plants because the gases escape from their surface in bubbles that can be easily seen and collected.

Investigating oxygen production in plants

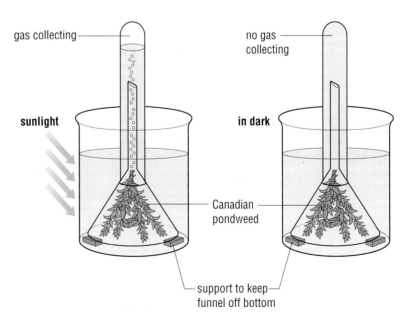

Figure 11.4 Apparatus for investigating oxygen production.

Two samples of Canadian pondweed were set up as shown in Figure 11.4. One was put in a sunny place and the other was kept in the dark. After about a week the amount of gas collected in each test-tube was examined. The plants in the dark had not produced any gas. The plants in the light had produced gas and when it was tested with a glowing splint the splint relighted, showing that the gas contained more oxygen than normal air.

Revising the work so far

From the result of this experiment the equation can be further modified to:

carbon dioxide + water → carbohydrate + oxygen

Moving on

Having established that the carbohydrate starch is formed in leaves it may seem reasonable to find out what affects the presence of starch in leaves. What is it in the plant that allows the reaction to happen? Does the reaction happen all the time or only at certain times of the day or night?

Testing the effect of light on a destarched plant

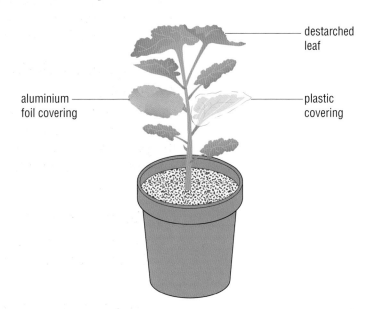

Figure 11.5   Two leaves of a destarched plant covered with plastic and aluminium.

Two leaves of a destarched plant were set up as shown in Figure 11.5 and left for over four hours in daylight. After that time they were removed and tested for the presence of starch. The leaf kept in the transparent plastic sheet contained starch. The leaf kept in the aluminium sheet did not contain any starch. This suggested that light is needed for starch to form in a leaf.

Revising the work so far

After reviewing the result of the effect of light on the leaf the equation can be modified to:

$$\text{carbon dioxide} + \text{water} \xrightarrow{\text{light}} \text{carbohydrate} + \text{oxygen}$$

Light provides the energy for the chemical reaction to take place. Some of the energy is stored as chemical energy in the carbohydrate.

Moving on

Having discovered a connection between the leaf, light and starch production it may seem reasonable to find out which part of the leaf is important. As most leaves are green it may be suggested that the green pigment, chlorophyll, which is found in chloroplasts of the leaf, is important. If it is lacking, starch should not be made. This hypothesis can be tested by using a variegated leaf, which has some cells that do not have chlorophyll thus making parts of the leaf appear white.

Investigating chlorophyll and starch production

A destarched variegated plant was left in daylight for over four hours. A leaf was then removed and tested for starch. The parts that were green contained starch but the parts that were white did not contain starch. This suggested that chlorophyll is needed for the leaf to produce starch.

Revising the work so far

After reviewing the result of the effect of chlorophyll on starch production, the equation can be modified to:

$$\text{carbon dioxide} + \text{water} \overset{\text{light}}{\underset{\text{chlorophyll}}{\longrightarrow}} \text{carbohydrate} + \text{oxygen}$$

Figure 11.6 A variegated pelargonium called Lady Plymouth.

Further experiments showed that the carbohydrate starch was built up in stages from subunits of a substance called glucose. The equation for starch production, or photosynthesis, is now written as:

7 Try to describe photosynthesis in your own words.

light

carbon dioxide + water → glucose + oxygen

chlorophyll

Photosynthesis is a chemical reaction. Chemical formulae can be used in the equation instead of words to make a symbol equation:

$$6CO_2 + 6H_2O \rightarrow C_6H_{12}O_6 + 6O_2$$

Fate of glucose

Glucose may be used to release energy in the process of respiration. The energy released is used for all life processes in plant cells. Glucose is also used to make many other molecules in the plant. It may be used to make cellulose for the cell walls or turned into fats for the cell membranes. Glucose may be changed into starch, which is an energy store for the plant, or made into sugars in fruits so that their sweet taste makes them attractive to animals (see page 23). Nitrogen and sulphur join with the elements in glucose to make amino acids and proteins.

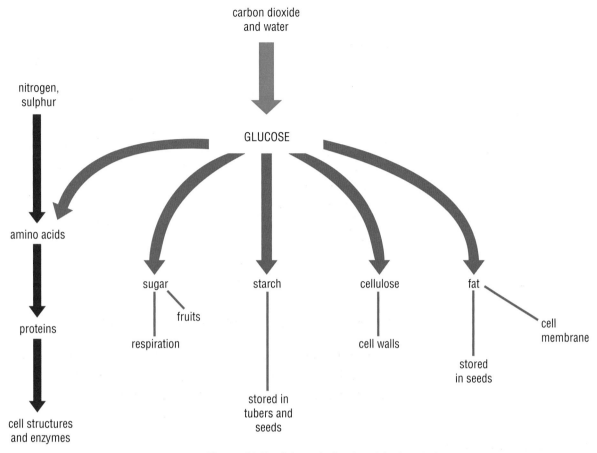

Figure 11.7 Schematic drawing of the fate of glucose.

8 What do plants take from the air and give to the air when they respire?

9 What do plants take from the air and give to the air during photosynthesis?

10 Compare the equations for photosynthesis and respiration.

11 How does the amount of
   a) carbon dioxide and
   b) oxygen vary around a plant over a 24-hour period? Explain your answer.

Plant respiration

Plant cells need energy to drive their life processes. As in animals, this energy is released in respiration (see page 101):

glucose + oxygen → carbon dioxide + water + energy

Plants respire 24 hours a day. They take in oxygen and produce carbon dioxide. During daylight photosynthesis also takes place. In this reaction carbon dioxide is used up and oxygen is produced. In bright sunlight the speed at which plants produce oxygen is greater than the speed at which they use up oxygen in respiration.

Glucose and starch

Glucose is soluble in the cell sap. If the concentration of glucose in the cell sap is too high, too much water is drawn into the cell. When glucose is made in large quantities in the leaf cells it is converted into starch, which is insoluble and does not affect the way water enters or leaves the plant cells. When the concentration of glucose in the cell becomes low the starch is converted back into glucose.

The food in plants

12 When will the amount of glucose in a leaf cell rise to a high concentration? Explain your answer.

13 Why does starch form?

14 When will the amount of starch in a leaf cell decrease? Explain why this happens.

Plants make glucose in the process of photosynthesis. This can then be used by the plant in a number of different ways. As already described, it may be turned into starch and stored to provide energy for later use. It may also be turned into fat and stored in some seeds to provide energy later. It may be changed into cellulose to make cell walls, or combined with minerals to make cell structures and enzymes. Minerals are vital for the health of a plant, and are taken into the plant as mineral salts.

Mineral salts

When chemists began studying plants they discovered that they contained a wide range of elements. With the exceptions of carbon, hydrogen and oxygen the plants obtained these elements from mineral salts in the soil.

The importance of each element was assessed by setting up experiments in which the plants received all the necessary mineral salts except for the one under investigation. From these studies it was found that:

15 Put the information about mineral salts and their uses by plants into a table. Include information about what happens if the mineral salt is missing.

16 What mineral might be missing if the leaves go yellow?

17 Why might a plant show poor growth?

- nitrogen is taken in as nitrates and is needed to form proteins and chlorophyll. Without nitrogen the plant's leaves turn yellow and the plant shows poor growth.
- phosphorus is taken in as phosphates and is needed to make chemicals for the transfer of energy in photosynthesis and respiration. Without phosphorus a plant shows poor growth.
- potassium is taken in in potassium salt and helps the plant to make protein and chlorophyll. If it is lacking the leaves become yellow and grow abnormally.

Figure 11.8   Plants showing mineral deficiency.

Path of minerals through living things

When animals eat plants they take in the minerals and use them in their bodies. Some of the minerals are released in the solid and liquid wastes that animals produce. As bacteria feed on these wastes, the mineral salts are released back into the soil. The mineral salts are also released when the plants and animals die and microbes break down their bodies in the process of feeding. Plants, animals and their wastes are biodegradable. This means they can be broken down into simple substances that can be used again to make new living organisms. These simple substances have been recycled since the beginning of life on Earth.

18 Could you be a recycled dinosaur? Explain your answer.

19 If a plant is over-watered all the spaces between the soil particles become filled with water. How does this water-logged soil affect
a) the plant roots and
b) the plant's growth?

Water and minerals in the plant

Most plant roots have projections called root hairs. The tips of the root hairs grow out into the spaces between the soil particles. There may be up to 500 root hairs in a square centimetre of root surface. They greatly increase the surface area of the root so that large quantities of water can pass through them into the plant. The water in the soil is drawn into the plant to replace the water that is lost through evaporation from the leaves. The plant does not have to use energy to take the water in.

Mineral salts are dissolved in the soil water. The plant has to use energy to take them in. This energy is provided by the root cells when they use oxygen in respiration. The roots get the oxygen from the air spaces between the soil particles.

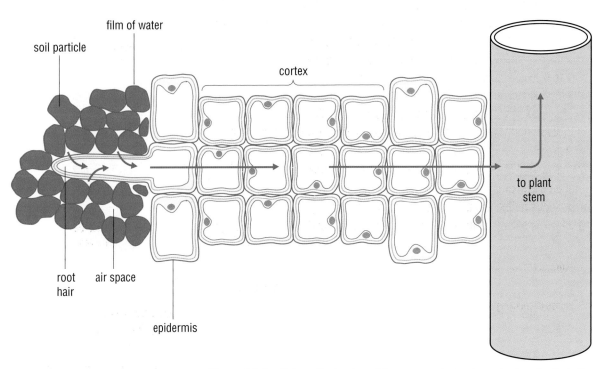

Figure 11.9 Schematic drawing of the movement of water and mineral salts in the root of a plant.

Carnivorous plants

Some plants live in conditions where minerals are unavailable. They are therefore unable to take up minerals from the soil. These plants have developed a way of getting the elements they need. Their leaves have adapted to allow them to trap and kill animals. The most important element required is nitrogen.

Butterwort and sundew grow in peaty bogs in the United Kingdom. The Venus fly trap grows in peaty bogs and waterlogged ground in the south-east of the United States. Pitcher plants (see Figure B) are found in tropical rainforests, in marshes in the United States and in some swamps in Australia. Bladderwort (see Figure B) is found in ponds in the United Kingdom.

The butterwort has leaves arranged in a rosette around the flower stalks. Each leaf is from 2 to 8 cm long. On the upper surface of the leaf are hairs that secrete a liquid containing protein-digesting enzymes. The edges of the leaf turn up to make a lip that prevents the sticky liquid from flowing away into the soil. If an insect lands on the liquid it cannot escape. Its soft parts are digested and absorbed into the leaf.

The leaves of the sundew have long stalks and circular blades. The upper surface of the leaf blade is covered in hairs that secrete sticky drops of a liquid that contains protein-digesting enzymes. When an insect lands on the leaf it sticks to the hairs and the enzymes digest its soft parts, leaving the hard parts to be blown away by the wind.

Each leaf of the Venus fly trap is divided into two halves that can spring together in 0.03 seconds. There are three hairs on each half of the leaf which act as triggers. If an insect lands on the leaf and touches them the trap is sprung. The spines on the edges of the leaf interlock and

Butterwort

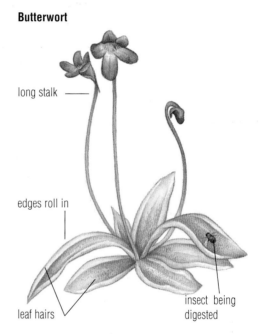

long stalk

edges roll in

leaf hairs

insect being digested

Venus fly trap

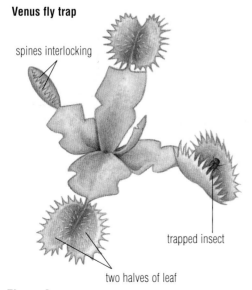

spines interlocking

trapped insect

two halves of leaf

Figure A

Sundrew

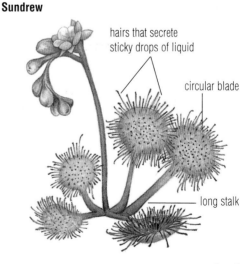

hairs that secrete sticky drops of liquid

circular blade

long stalk

(continued)

stop the insect escaping. A liquid is secreted from the leaf's surface which digests and absorbs the insect's body. In 24 hours the leaf opens again. After it has digested four insects the leaf dies and is replaced by a new one.

The rainforest pitcher plants are able to climb trees because they have cords, called tendrils, that stick out of their leaves. They curl around twigs and branches and give the pitcher plant support. The end of a tendril forms a hollow tube with a lid. This tube is called a pitcher and when it is fully grown the lid opens and water collects in it. The largest pitchers can hold up to 2 litres of water. There are glands at the mouth of the pitcher which produce nectar to attract insects. The inside walls of the pitcher are very smooth so that the insect loses its grip and falls into the water. Some pitcher plants also produce a drug in their nectar which makes the insect lose co-ordination and fall into the pitcher. There are hairs on the pitcher walls that point downwards to prevent the insect escaping. The insect drowns. As its body decays it releases nutrients into the water. These are absorbed by the walls of the pitcher. The pitcher plants that grow in the United States secrete enzymes and acids to digest the insects. In the Australian species, enzymes and bacteria break down the insects' bodies.

The bladderwort is an aquatic plant that has pin-head sized traps on its feathery leaves. Each trap is called a bladder. The bladder is a globe-shaped structure that has a cavity which can be filled with water or air. The trap is set by removing the water from the cavity. Inside each bladder are cells that absorb water. At the bladder's mouth is a trap door, trigger hairs and cells that contain sugar to attract prey. When a water flea touches the trigger hairs the trap door opens and water rushes into the bladder carrying the water flea with it. The door closes behind it and the water flea is absorbed within a few days.

1 What kind of conditions do not provide enough minerals for plants to grow well?
2 Why is nitrogen particularly important for plant growth?
3 How successful are the plant traps? Explain your answer.
4 What methods are used by carnivorous plants to break down an insect's body?

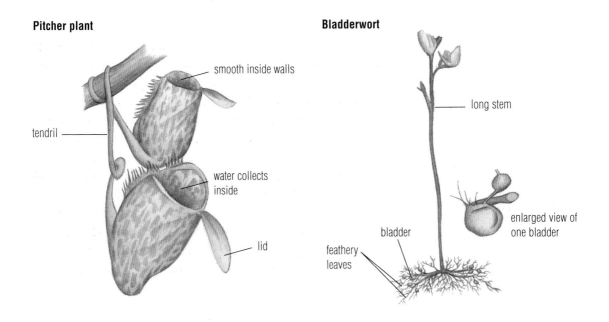

Pitcher plant

smooth inside walls

tendril

water collects inside

lid

Bladderwort

long stem

enlarged view of one bladder

bladder

feathery leaves

Figure B

Oxygen and carbon dioxide in the atmosphere

About 20% of the atmosphere is composed of oxygen and about 0.03% is composed of carbon dioxide. These two amounts remain the same from year to year. The reason they do not change is that the carbon dioxide produced by animals and plants in respiration is used up in photosynthesis, and the oxygen produced by the plants is used up by plants and animals in respiration.

20 Why do humans not suffocate at night when the plants around them cannot photosynthesise?

21 What effect will reducing the number of plants on the surface of our planet have on the animals?

Figure 11.10 Some of the oxygen that these deer are breathing has been produced by the trees around them.

Carbon cycle

The carbon dioxide taken into a plant is used to make glucose, which may be transported to a storage organ and converted into starch. If the storage organ is a potato, for example, it may be dug up out of the ground, cooked and eaten. The starch is broken down in digestion to glucose and taken into the blood. In the body the glucose may be used for respiration and the carbon is released as carbon dioxide. If too much high-energy food is being eaten the glucose may be converted into fat and the carbon remains in the body. When the body dies, microbes feed on it and break it down into simple substances. The microbes thus release the carbon back into the air, as carbon dioxide, when they respire.

22 Re-read the account of the carbon cycle and draw the paths that carbon can take.

23 Now add in the path that the carbon would take if the potato plant died.

24 Why is the path called the carbon cycle?

◆ SUMMARY ◆

◆ Iodine solution is used in the test for starch (*see page 192*).
◆ Boiling water and ethanol are used, with care, to remove chlorophyll from a leaf (*see page 193*).
◆ A plant is destarched by leaving it in the dark for 2 or 3 days (*see page 194*).
◆ Water and carbon dioxide are the raw materials of photosynthesis (*see page 195*).
◆ Light and chlorophyll are needed for a plant to photosynthesise (*see page 197*).
◆ Carbohydrate (glucose and starch) and oxygen are the products of photosynthesis (*see page 198*).
◆ Mineral salts are needed for healthy plant growth (*see page 200*).
◆ Water and minerals enter the plant through the root hairs (*see page 202*).
◆ Photosynthesis and respiration keep the levels of oxygen and carbon dioxide in the air constant (*see page 205*).
◆ Carbon passes from the air to a plant, then to an animal and finally a microbe releases it into the air again as it moves around the carbon cycle (*see page 205*).

End of chapter questions

How does a growing cucumber's weight change in a day? Large cucumbers appear to grow quickly. In this experiment a cucumber plant was placed close to a top-pan balance and one of its growing cucumbers was placed on the pan. The weight of the cucumber was measured every hour between 9.00 am and 4.00 pm. The weight was displayed in the graph in Figure 11.11.

For discussion
The rainforests have been described as the world's lungs. What do you think this means?

1 What was the gain in weight over the seven-hour period?
2 Construct a table to display the increase in weight in each of the seven hours.
3 When was the period of
   a) greatest and
   b) least growth?
4 If the cucumber plant had some of its leaves removed before the experiment, how would you expect its growth graph to compare with the graph in this experiment? Explain your answer.

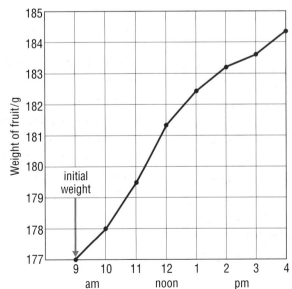

Figure 11.11

12 Plants for food

You may start your day with a breakfast like this one.

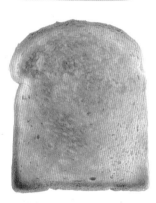

corn flakes

golden, crisp, light
flakes of corn

500 gram ℮

Figure 12.1 Examples of some breakfast foods.

1 Write a food chain for each item in this meal.
2 Write down the items in your breakfast. Make a food chain from each of these items. How many food chains feature animals as well as plants?
3 Read through Chapter 5 again and assess the healthiness of your breakfast. If it is unhealthy what could you do to change it?

Where do all these foods come from? The cereal flakes are made from grains of maize – the grains you also eat as corn on the cob. The milk comes from a cow, but what does the cow eat? (see Figure 12.2 for the answer). The slice of bread is made from the flour of another cereal called wheat. It is spread with a margarine made from sunflower oil, and honey that is made from the nectar of flowers by bees.

What parts of a plant are eaten?

Seeds and fruits

If you look at Figure 1.23 on page 19 you can see that an ovule is surrounded by an ovary wall, and a pollen tube nucleus is about to join with an egg cell nucleus. When the two nuclei join together, fertilisation takes place and the ovule forms a seed while the ovary wall becomes the fruit, both of which are often eaten.

Cereals

Cereals are crops that have been selectively bred from species of grass plants. A cereal grain is a fruit containing a single seed. The fruit forms a thin, dry coat around the seed. In wheat the fruit is called the bran. The fruit is usually removed from the grain before the grain is used to make food.

The main cereal crops grown in temperate climates are:

- wheat – ground up to make flour for a wide range of food products
- maize – for breakfast cereals, sweet corn, corn on the cob
- barley – used in soups, stews and for farm animal food
- oats – used for making porridge and for farm animal food
- rye – used to make certain types of bread and biscuits.

The main cereal crops grown in tropical climates are:

- rice – simply boiled or fried, or made into rice pudding
- millet – mixed with water to make a type of porridge.

Legumes

4 Use Table 5.2 on page 86 to answer this question.
   a) Compare the nourishment provided by peas and lentils.
   b) Could either be used as a substitute for chicken? Explain your answer.

The main crops that produce seeds for food are the legumes. Peas, beans and lentils are seeds and are sometimes called pulses. They are produced in a fruit called a pod. In some kinds of legumes, such as runner beans and mangetout, the pod and the seeds are eaten. The pea nut is an unusual legume. After fertilisation the fruit and seeds develop under the soil. The fruit forms a woody shell around the seeds that we call peanuts.

Nuts

A true nut is made from a fruit which has developed into a woody wall around the single seed. The brazil nut is an example of a nut with a very strong woody wall which must be cracked to release the seed – the part we call the brazil nut.

Fruits

In many plants after fertilisation the ovary wall swells up and becomes fleshy or succulent. Its surface may become a bright colour to make the fruit look attractive enough to eat. Some fruits such as the plum or cherry contain just one seed which has a tough woody coat. This seed is known as the stone. Other fruits such as the orange and the tomato have a centre which is full of small seeds. The apple and the pear are not really fruits and are known as false fruits. The fleshy part which makes the fruit does not form from the ovary wall, but forms from the top of the flower stalk which swells up and grows round the ovary and its seeds.

Vegetables

In everyday life beans, lentils, peas and tomatoes are all called vegetables, but there are many other plants in this group of foods. Each vegetable plant is grown so that a particular part of its body can supply nourishment.

Table 12.1 shows the parts of some vegetables that are used as food.

5 Which parts of the following plants are used as food – radish, leek, parsnip, cress, cucumber, chicory, kohlrabi?

Table 12.1

Vegetable	Part of plant
broccoli, cauliflower	flower
aparagus	stem
Brussels sprout	bud
celery	leaf stalk
lettuce, onion	leaves
potato	tuber
carrot, beetroot, cassava, yam	root

The food of farm animals

There are three groups of food fed to farm animals. They are fibrous foods (grass, hay, silage, lucerne and straw), succulent foods (beet, kale and mangolds), and concentrated foods made from cereals, fishmeal, oil cake and pulses. The proportions of the food groups in the diet of four kinds of farm animals are shown in Figure 12.2.

6 How does the diet of poultry compare with the diets of cattle, sheep and pigs?

7 Which animal group can provide food without being slaughtered? Explain your answer.

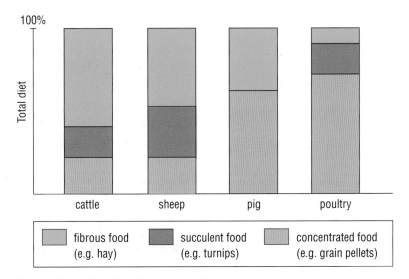

Figure 12.2   The diets of four kinds of farm animal.

BSE

BSE stands for bovine spongiform encephalitis. It is popularly known as 'mad cow disease', and is believed to have developed in the following way. Part of the diet of cattle is made from concentrated food (see Figure 12.2). In the manufacture of some concentrated food the body parts of animals which were not to be sold as food were taken from the abattoir and made into pellets for cattle to eat. At some time sheep carcasses, which were infected with a disease called scrapie, were sent for processing and were made into food. The infection persisted in the carcasses after processing and caused the development of BSE in cattle between three and five years old. BSE is a fatal disease, which develops slowly in the brain and nervous system.

A disease which is similar to BSE also develops in humans. It is called Creutzfeldt–Jakob Disease. Some medical experts believe that eating meat containing the nervous tissue of a BSE-infected animal can cause BSE in humans.

1 What is your reaction to the practice of feeding cattle on the bodies of other animals? Explain your answer.

2 What do you think was done to reduce the alleged risk of people becoming infected with BSE?

3 Ideas relating to health and disease can change quite quickly. What is the current view on BSE?

8 What kind of fertiliser gives you the greater control over providing minerals for a crop? Explain your answer.

Fertilisers

When plants are grown for food they are eventually harvested and taken out of the soil. This means that the minerals they have taken from the soil go with them to market, and there are therefore fewer minerals left in the soil for the next crop. Fertilisers are added to the soil to replace the minerals that have been taken away in the crop. They are also added in quantities that will make sure that the plants grow as healthily as possible and produce a large crop. The amount of fertiliser added to the soil has to be carefully calculated. If too little is added, the amount of food produced by the crop (called the yield) will be small. If too much is added, the plants will not use all the minerals and they may be washed into streams and rivers and cause pollution (see page 153). There are two kinds of fertiliser. Inorganic fertilisers such as ammonium nitrate are manufactured chemical compounds (see *Chemistry Now! 11–14* second edition Chapter 13). Organic fertilisers are made from the wastes of farm animals (manure) and humans (sewage sludge).

Inorganic fertilisers can give crops an almost instant supply of minerals, as the minerals dissolve in the soil water as soon as they reach it, and can be taken up by the roots straight away. The minerals in manure are realeased more slowly as decomposers (see page 147) in the soil break it down.

Inorganic fertilisers are light in weight and so can be spread from aeroplanes and helicopters flying over the crops. This means that they can be applied to the crop at any time without damaging the crop. Manure is too heavy to be spread in this way and must be spread from the trailer attached to the tractor. If the manure was spread while the crop was growing the tractor and trailer would damage the crop, so the manure must instead be spread on the soil before the crop is sown. This also allows some time for the manure to release minerals into the soil.

Fertilisers and soil structure

A good soil has particles of rock bound together with humus (decomposed plants and leaves) to form lumps called crumbs. The crumbs do not interlock but settle on each other loosely, with air spaces between them. The air provides a source of oxygen for the plant roots and for organisms living in the soil. The spaces also allow the roots to grow easily through the soil. The humus acts like a sponge and holds onto some of the water that passes through the soil. The plant roots draw on the water stored in the humus.

9 If a farmer uses only inorganic fertilisers how may the soil organisms be affected?

In time humus rots away, and in natural habitats it is replaced by the decaying bodies of other organisms. When manure is added to the soil it adds humus and this helps to bind the rock particles together and keep the soil crumbs large. If the soil receives only inorganic fertilisers the humus is gradually lost and the soil crumbs break down. The rocky fragments that remain form a dust which can be easily blown away by the wind.

Pests and pesticides

Fertilisers are used to make the crop yield as high as possible. The crop plants may however be affected by other organisms, which either compete with them for resources, or feed on them. These organisims are known as pests, and chemicals called pesticides have been developed to kill them. There are three kinds of pesticides – herbicides, fungicides and insecticides.

The problem with weeds

When a crop is sown the seeds are planted so that the plants will grow a certain distance apart from each other. This distance allows each plant to receive all the sunlight, water and minerals that it needs to grow healthily and produce a large yield.

A weed is a plant growing in the wrong place. For example, poppies may be grown in a flower bed to make a garden look attractive, but if they grow in a field of wheat they are weeds because they should not be growing there. Weeds grow in the spaces between the crop plants and compete with them for sunlight, water and minerals. This means that the crop plants may receive less sunlight because the weeds shade them, and receive less water and minerals because the weed plants take in some for their own growth. Weeds can also be infested with microbes which can cause disease in the crop plant. For example, cereals may be attacked by fungi that live on grass plants growing as weeds in the crop.

Figure 12.3 Poppies growing as weeds in a field of wheat.

10 How does the use of herbicides on a farm affect the honey production of a local bee keeper? Explain your answer.

Herbicides

Weeds are killed by herbicides. There are two kinds of herbicide – non-selective and selective. A non-selective herbicide kills any plant. It can be used to clear areas of all plant life so that crops can be grown in the soil later. It must not be used when a crop is growing as it will kill the crop plants too. A selective herbicide kills only certain plants – the weeds – and leaves the crop plants unharmed. It can be used when the crop is growing.

Herbicides may be sprayed onto crops from the air. Some of the herbicide may drift away from the field and into surrounding natural habitats. When this happens, the herbicide can kill plants there. Many wild flowers have been destroyed in this way.

Fungicides

A fungicide is a substance that kills fungi. Fungal spores may be in the soil of a crop field or floating in the air above it. They may even settle on the seeds before they are sown. Fungicides are coated on seeds to protect them when they germinate. They are also applied to the soil to prevent fungi attacking the roots, and are sprayed on crop plants to give them a protective coat against fungal spores in the air.

Insecticides

When a large number of plants of the same kind are grown together they can provide a huge feeding area for insects. Large populations of insects can build up on the plants and cause great damage. Insecticides are used to kill these insect pests. There are two kinds of insecticides – narrow spectrum and broad spectrum insecticides. Narrow spectrum insecticides are designed to kill only certain kinds of insects, and leave others unharmed. Broad spectrum insecticides in contrast kill a wide range of insects – not only those that feed on the crop, but also predatory insects which may prey on them. If insecticides drift away from the fields after spraying, they can kill other insects in their natural habitats. They can also move along the food chain.

11 You have a field of weeds and need to plant a crop in it. How could you use pesticides to prepare the ground for your crop and protect it while it is growing?

A poison in the food chain

In 1935 Paul Müller (1899–1965) set up a research programme to find a substance that would kill insects but would not harm other animals. Insects were his target because some species are plant pests and devastate farm crops and others carry microbes that cause disease in humans. The substance also had to be cheap to make and not have an unpleasant smell. In 1939 he tried a chemical called dichlorodiphenyltrichloroethane, shortened to DDT, that was first made in 1873. DDT seemed to meet all his requirements and soon it was being made in large amounts and used worldwide.

In time, some animals at the end of the food chains (the top carnivores) in the habitats where DDT had been sprayed to kill insects were found dead. The concentration of the DDT applied to the insects was much too weak to kill the top carnivores directly so investigations into the food chains had to be made.

In Clear Lake, California, DDT had been sprayed onto the water to kill gnat larvae. The concentration of DDT in the water was only 0.015 parts per million (ppm), but the concentration in the dead bodies of the grebes (fish-eating water birds) was 1600 ppm. When the planktonic organisms were examined their bodies contained 5 ppm and the small fish that fed on them contained 10 ppm.

It was discovered that DDT did not break down in the environment but was taken into living tissue and stayed there. As the plankton in the lake were eaten by the fish the DDT was taken into the fishes' bodies and built up after every meal. The small fish were eaten by larger fish in which the DDT formed higher concentrations still. The grebes ate the large fish and with every meal increased the amount of DDT in their bodies until it killed them.

In Britain, the peregrine falcon is a top carnivore in a food chain in moorland habitats, although it visits other habitats outside the breeding season. The concentration of DDT in the bodies of the female falcons caused them to lay eggs with weak shells. When the parents incubated the eggs their weight broke the shells and the embryos in the eggs died.

Alerting the world to pesticide danger

Figure A Rachel Carson.

Rachel Carson (1907–1964) was very interested in the plants and animals around her home when she was a child. When she went to college she had plans to become a writer, but took a degree in science and then a Masters degree in zoology. She satisfied her interest in writing by preparing the scripts for radio programmes for the United States Bureau of Fisheries. She continued her work in science with the federal service working as a marine biologist, and used her literary skills in the post of editor-in-chief of all publications for the United States Fish and Wildlife Service.

Carson's first book, *Under the Sea Wind*, was published in 1941. Although it was well written it did not become well known. Ten years later her second book, *The Sea Around Us*, was published and was so successful that is was translated into thirty-two languages. After this success Carson became a full-time writer, and in 1962 her book *Silent Spring* was published. In it she showed how living things in their environment were related to each other, and how the indiscriminate use of pesticides could lead to the destruction of many forms of life, not just the pests that farmers wished to kill.

The title *Silent Spring* refers to the death of song birds due to eating insects poisoned with pesticides. *Silent Spring* made the gereral public aware of the dangers of the excessive use of pesticides. In turn people demanded to know more about the effects of pesticides, and scientists began research programmes to investigate them. As a result of their work many countries now have laws which control the use of pesticides so that the wildlife is protected.

Scientists who can write well can hold the attention of others and inspire them. Here is a short extract from *Silent Spring*.

Those who contemplate the beauty of the Earth find reserves of strength that will endure as long as life lasts. There is a symbolic as well as actual beauty in the migration of birds, the ebb and flow of tides, the folded bud ready for spring. There is something infinitely healing in the repeated refrains of nature – the assurance that dawn comes after night and spring after winter.

(From *Silent Spring* by Rachel Carson, first published by Houghton Mifflin, 1962.)

1 How did Rachel Carson's childhood interests shape her career?

2 How did she combine her interest in writing with her work in science?

3 What might have happened if *Silent Spring* had not been written?

4 What is a refrain?

5 How is the migration of birds like the ebb and flow of the tides?

6 What other refrains of nature can you think of?

7 Select something you have studied in this book that you particularly liked and write about it as if you were trying to interest others.

The perfect environment for growth

When plants are grown for food, every attempt is made to ensure that the crop yield is as large as possible. Applying fertilisers helps crop growth, and the use of pesticides keeps other organisms from damaging the crop. There are however other factors that affect growth. They are the factors which affect photosynthesis – light, temperature and the amount of carbon dioxide in the air (see Chapter 11).

The effect of light

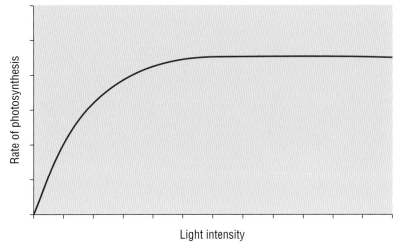

Figure 12.4   The effect on light intensity on the rate of photosynthesis.

Figure 12.4 shows the results of measuring the rate of photosynthesis at different light intensities. When the light intensity is increased from darkness, the rate of photosynthesis rises steadily. At a certain intensity it stops increasing and remains steady, even though the intensity is increased much more.

The effect of temperature

When plants are kept in a high light intensity but the temperature of the surroundings is changed, the rate of photosynthesis changes as shown in Figure 12.5.

The graph can be used to work out the best temperature for maximum photosynthesis.

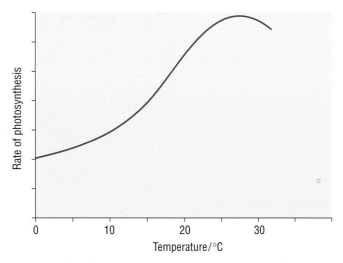

Figure 12.5 The effect of temperature on the rate of photosynthesis.

The effect of carbon dioxide

Figure 12.6 shows the results from the following experiment. One set of plants was set up at 25°C in air with a low concentration of carbon dioxide. Another set of plants was set up at 25°C in air with a high concentration of carbon dioxide. When the light intensity was gradually increased the rate of photosynthesis of the plants in each set was measured.

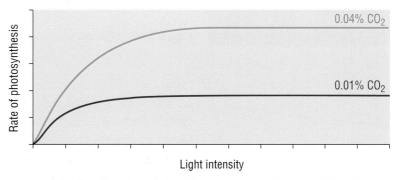

Figure 12.6 The effect of carbon dioxide concentration on the rate of photosynthesis.

Combining the three factors

The results of the experiments show that the amount of light, the temperature and the amount of carbon dioxide in the air all affect the rate of photosynthesis. This implies that the rate of photosynthesis can be increased to a maximum by carefully regulating these three factors in the environment of the plant. These factors cannot be controlled in an open field but they can be controlled by using a glasshouse.

The glasshouse

For a large part of the year the plants in a glasshouse receive all the light they need from the Sun. The growing season can be extended by providing extra light from special lamps. Plants use certain wavelengths of sunlight for photosynthesis and these lamps emit light at these wavelengths.

Figure 12.7 A glasshouse with blinds and ventilators.

For discussion

It has been proposed to construct a huge glasshouse at a farm close to a village. How do you think the villagers will receive the news?

14 How do you think the action of
a) the blinds and
b) the ventilators are controlled?

In summer a glasshouse may receive all the heat it needs from the Sun too. In winter, heaters can be used to lengthen the growing season. The amount of carbon dioxide in the air in a glasshouse can be increased by pumping in extra carbon dioxide. In winter, paraffin heaters can be used to supply extra carbon dioxide to the air as they warm it up.

While a crop is growing the conditions need to be kept as constant as possible. Very high light intensities from the Sun on a clear summer day can damage chlorophyll, and this in turn reduces the rate of photosynthesis. This can be prevented by having blinds which extend automatically when the light intensity increases to a damaging level. If the glasshouse becomes too hot ventilators can open automatically to allow the hot air to escape and cooler air to enter.

GM foods

In addition to working out the best way to grow crops, scientists also look at ways to improve the crop plants. One way to do this is by selective breeding (see page 168). A second way is by genetic modification. Food produced by plants which have been genetically modified are called GM foods. A plant is genetically modified by having genes from another organism added to it. The addition of genes may help it to grow better or help the food to be processed more cheaply. For example, some tomato plants were modified so that they produced tomatoes that made a thick tomato puree more cheaply than other kinds of tomatoes.

Genetically modified organisms are new organisms. Scientists disagree on their use because if they enter a natural habitat they may breed with organisms there to produce a more unusual organism.

1 How is genetic modification different from selective breeding?

2 Ideas and practices relating to genetically modified food can change quite quickly. What are the current views and practices on genetically modified foods?

Organic farming

In the 18th Century Charles Townsend (1674–1738) introduced the idea of crop rotation into the growing of food plants. Figure 12.8 shows a plan for crop rotation. Each crop has particular pests. If the crop is grown every year in the same field the population of pests builds up and damages the yield. By rotating the crop the populations are not allowed to build up. One of the crops is used for grazing animals. As they spend time in the field they provide manure to build up the minerals that are used by the other crops in the rotation scheme. Each crop will grow well even if the soil it is growing in has not been manured for three years, because each crop takes a different range of minerals from the soil.

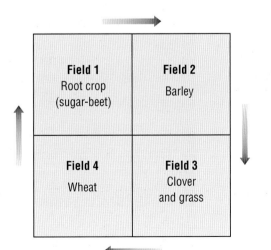

The arrows show the movement of crops in the following year.
In year 2 the root crop will be planted in field 2.

15 Construct plans for the four fields in Figure 12.8 for the next three years.

Figure 12.8   Crop rotation plan.

16 Organic food is more expensive than other food. Why?

17 Should all farms be organic farms? Explain your answer.

In the 20th Century it was found that crops could be produced more cheaply if the hedges were removed to make very large fields, and if only one crop was grown. This allowed sowing and harvesting machines to work more easily. The use of inorganic fertilisers meant that the minerals could be given to the plants as they needed them and pesticides could be used to kill off any organisms that threatened to damage the crops. However, there are fears that chemicals used in growing these crops are left on the food and can cause illnesses in people. These fears have meant that some farmers now farm using the old crop rotation system, and use methods which do not involve chemicals. For example, manure is used instead of inorganic fertilisers, weeds are removed by hand instead of by the application of herbicides, and plants which keep insects away are planted among the crop plants instead of the farmer using insecticides.

Figure 12.9 Crops being farmed organically.

◆ SUMMARY ◆

♦ Plants provide us with food in the form of seeds, cereal grains, nuts, fruits and vegetables (*see pages 208–209*).

♦ Plants provide food for farm animals (*see page 210*).

♦ Fertilisers provide minerals for plant growth (*see page 211*).

♦ Pesticides remove organisms that can damage crops (*see pages 212–213*).

♦ Poisons can move through food chains (*see page 214*).

♦ Plant growth can be controlled by regulating factors such as the amount of light, heat or carbon around a plant (*see pages 216–218*).

♦ Organic farming does not use pesticides or inorganic fertilisers (*see page 219*).

End of chapter questions

1 a) Write down twelve foods that you frequently eat.

b) Work out a food chain for each one.

c) On which food plants does your diet greatly depend?

For discussion

Imagine that you are going to live on an island with about two dozen other people. The island has four seasons, with a warm summer and a cool winter. It has a hill near its centre which provides some shelter from the cold prevailing wind. Much of the low land on the sheltered side of the hill is woodland. There is a small population of fish in the sea around the island which can provide a little food. Work out a plan for providing food for yourself and the rest of the group for a stay of two years on the island. You may take any plants, animals, pesticides, fertilisers and building materials that you wish, and you may begin your stay on the island in any season.

Glossary

A

adaptation   The way a living thing is suited to its habitat so that it can survive there. Adaptation can also mean the process by which living things become more suited to their habitat.

addiction   A condition in which a person is unable to lead a normal life without taking drugs or alcohol on a regular basis.

adolescence   The time in a person's life when they change from a child to an adult.

aerobic respiration   The release of energy from food using oxygen.

alimentary canal   The digestive tube that begins with the mouth and ends with the anus. It is also sometimes called the gut.

amino acid   A molecule containing carbon, hydrogen, oxygen and nitrogen. It links up with other amino acids to form long-chain molecules called proteins.

amnion   A sac that surrounds the embryo that is filled with a watery fluid.

anaemia   An unhealthy condition that may be due to the lack of iron in the diet. One of the symptoms is tiredness.

anaerobic respiration   The release of energy from food without the use of oxygen.

antagonistic muscles   A pair of muscles in which each of the contracting muscles brings about a movement that is opposite in direction to the other.

anther   The organ in a flower that produces pollen grains.

antibiotic   A chemical made by some microbes or produced artificially by chemical reactions that is used to kill certain kinds of disease-causing bacteria.

antibody   A chemical made by some white blood cells to protect the body from disease-causing microbes and their toxins.

antigen   A feature found on the body of a disease-carrying microbe that stimulates the human body to produce antibodies.

artery   A blood vessel with elastic walls that carries blood away from the heart.

asexual reproduction   The process of producing offspring without the making of gametes and the process of fertilisation.

B

bile   A substance made by the liver and stored in the gall bladder. It is released onto food in the duodenum to aid the digestion of fats.

biodegradable   The property of a complex substance that allows it to be broken down into simple substances by the action of decomposers.

biomass   The mass of an organism or group of organisms after their bodies have been dried out.

biotechnology   The use of biological processes to make useful substances, such as antibiotics, and to produce new kinds of living organisms through genetic engineering.

C

calyx   The ring of sepals in a flower.

capillary   A blood vessel with one-cell thick walls through which substances pass between the blood and the surrounding cells.

carbohydrate   A nutrient made from carbon, hydrogen and oxygen. Most are made by plants.

carnivore   An animal that only eats other animals for food.

carpel   The female organ of a flower that produces the fruit and the seed.

cell (in biology)   The basic unit of life. The cell contains a nucleus, cytoplasm and membrane around the outside. The bodies of most living things are made from large numbers of cells.

chlorophyll   A green pigment found mainly in plant cells that traps energy from sunlight and makes it available for the process of photosynthesis.

chloroplast   A component of a cell. It is green and absorbs some of the energy of sunlight for use in photosynthesis.

chromosome   A thread-like structure that appears when the cell nucleus divides. It contains DNA.

cilia   Short hair-like projections that may form on the surface of a cell. They can beat to and fro to move the bodies of Protoctista or to help with the movement of fluids in animal systems.

clone   One of a number of identical individuals produced by asexual reproduction.

consumer   An animal that eats either plants or other animals.

corolla   The ring of petals in a flower.

cross-pollination   The transfer of pollen from the anthers of a flower on one plant to the stigma of a flower on another plant of the same species.

cytoplasm   A fluid-like substance in the cell in which processes take place to keep the cell alive.

D

digestion   The process of breaking down large food particles into small ones so that they can be absorbed by the body.

DNA (deoxyribonucleic acid)   A substance in the nuclei of cells that contains information, in the form of a code, about how an organism should develop and function.

E

ecology   The study of living things in their natural surroundings or habitat.

ecosystem   An ecological system in which the different species in a community react with each other and with the non-living environment. Ecosystems are found in all habitats such as lakes and woods.

egestion (*see also* excretion)   The release of undigested food and other contents of the alimentary canal from the anus.

embryo   The body of an organism in its early development from a zygote (*see also* zygote). In humans an embryo develops in the womb in the first two months of pregnancy (*see also* fetus).

endoskeleton   A skeleton on the inside of the body, as occurs in vertebrates.

enzyme   A chemical made by a cell that is used to speed up chemical reactions in life processes such as digestion and respiration.

evolution   The process by which one species of living thing is believed to change genetically over a period of time in order to develop into a more complex organism.

excretion (*see also* egestion)   The release of waste products made by chemical reactions inside the body.

exoskeleton   A skeleton on the outside of the body, as occurs in arthropods.

F

fats   Food substances that provide energy. They belong to a group of substances called lipids, which include oils and waxes.

fermentation   A type of anaerobic respiration that occurs in yeast and bacteria. Some fermentation processes are used to make alcohol.

fertilisation   The fusion of the nuclei from the male and female gametes that results in the formation of a zygote.

fetus   A stage in the development of the mammal in the womb when the main features of the animal have formed. In humans the fetus develops in the womb from the second to the ninth month of pregnancy.

fruit   A structure that forms from the ovary of a flowering plant after fertilisation has taken place.

G

gamete   A cell involved in sexual reproduction, i.e. a sperm or egg cell in animals.

gene   A section of DNA that contains the information about how a particular characteristic, such as hair colour or eye colour, can develop in the organism.

genetic engineering   The process of moving genes between different types of organisms to produce new organisms with particularly useful properties.

germination   The process in which the plant inside a seed begins to grow and bursts out of the seed coat.

gonadotrophin   A chemical produced by the pituitary gland that stimulates the reproductive organs of males and females to develop fully.

growth hormone   A chemical produced by the pituitary gland in the head. It makes the body grow.

H

habitat   The place where a particular living thing survives.

haemoglobin   The pigment in red blood cells that contains iron and transports oxygen around the body.

herbivore   An animal that eats only plants for food.

hormone   A chemical, secreted by a gland in the body, which travels in the blood and acts on particular parts of the body. It may produce changes in growth or activity.

hygiene   The study and practice of maintaining health by keeping the body clean.

I

immunisation   A process in which the body is made resistant or immune to a disease.

incubation   A process in which organisms, such as a developing chick in an egg or colonies of bacteria, are kept at a constant, raised temperature to aid their growth and development.

invertebrate   An animal that does not have a skeleton of cartilage or bone inside its body.

J

joint   A place where two bones meet. In movable joints the bones are held together by ligaments and are capped in cartilage to reduce friction.

L

lymphocyte   A white blood cell that makes antibodies to destroy bacteria.

M

menopause   The time in a woman's life, usually about the age of 50, when monthly periods (menstruation) stop.

menstruation   A period of time each month when the uterus loses the lining of its wall.

milk teeth   The first set of teeth that grows in the jaws of young mammals, including humans.

mineral (in biology)   A substance taken up from the soil water by the plant roots and used for growth and development of the plant. It is also an essential nutrient in the diet of animals.

mutation   A change in a gene or chromosome that occurs suddenly and produces a change in the development of the organism.

N

natural selection   The process by which evolution is thought to take place. Individuals in a species best suited to an environment will thrive there and produce more offspring, while less well suited individuals will produce fewer offspring. In time the less well suited will die out leaving the best suited individuals to form a new species.

nucleus (in biology)   The part of the cell that contains the DNA and controls the activities and development of the cell.

nutrient   A substance in a food that provides a living thing with material for growth, development and good health.

O

omnivore   An animal that eats both plants and animals for food.

organ   A part of the body, made from a group of cell tissues, that performs an important function in the life of the organism.

ovary   The organ where the female gametes are made in plants and animals.

P

peristalsis   The wave of muscular contraction that moves food along the alimentary canal.

phloem   A living tissue in a plant through which food made in the leaves passes to all parts of the plant.

photosynthesis   The process by which plants make carbohydrates and oxygen from water and carbon dioxide, using the energy from light that has been trapped in chlorophyll.

pistil   A structure made from a group of carpels.

placenta   A disc of tissue that is connected to the uterus wall and supplies the baby with oxygen and food from the mother's blood and releases waste from the baby into the mother's circulatory system.

plankton   Very small (including microscopic) organisms that live near the water surface in large aquatic environments such as oceans and lakes.

pollen   Microscopic grains produced by the anther, which contain the male gamete for sexual reproduction in flowering plants.

pollination   The transfer of pollen from an anther to a stigma.

protein   A substance made from   amino acids. Proteins are used to   build many structures in the bodies of living things.

puberty   The time of body growth in humans during which the reproductive organs become fully developed.

R

respiration   The process in which energy is released from food.

S

saliva   A watery substance produced by glands in the mouth that makes food easier to swallow and begins the digestion of carbohydrates.

seed   A structure that forms from the ovule after fertilisation. It contains the embryo plant and a food store.

self-pollination   The transfer of pollen from the anthers to the stigma of the same flower.

sepal   A leaf-like structure that protects a flower when it is in bud.

sexual reproduction   The form of reproduction in which gametes are formed, fertilisation takes place and a zygote is formed.

spore   A reproductive structure, containing one or more reproductive cells, produced by fungi and plants, such as mosses and ferns, that do not produce seeds. They are also produced by bacteria so that they can survive harsh environmental conditions.

stamen   A structure in a flower composed of the anther and the filament.

stigma   The region on a carpel on which pollen grains are trapped.

T

testes   The male reproductive organs in animals. They produce sperm.

tissue   A structure made from large numbers of one type of cell.

toxins   Poisons produced by bacteria which cause disease.

U

urea   A chemical made when amino acids are broken down in the body to make a carbohydrate called glycogen. It is excreted by the kidneys.

ureter   The tube connecting the kidney to the bladder, through which urine flows.

urethra   The tube connecting the bladder to the outside, through which urine passes. In males sperm also pass along this tube.

urine   A watery solution that contains urea.

uterus   The female organ in which the embryo and fetus develop. It is also known as the womb in humans.

V

vaccine   A substance that promotes the production of antibodies to protect the body from certain diseases.

vacuole   A large cavity in a plant cell that is filled with a watery solution called cell sap. May also occur as small, fluid-filled cavities in some animal cells and some Protoctista.

variation   A feature that varies among individuals of the same species, such as height or hair colour.

vein   A thin-walled blood vessel that transports blood towards the heart.

vertebrate   An animal that has a skeleton inside its body made of cartilage or bone.

vitamin   A substance made by plants and animals that is an essential component of the diet to keep the body in good health.

X

xylem   The non-living tissue in a plant through which the water and minerals pass from the root through to the shoot.

Z

zygote   The cell produced after fertilisation has occurred. It divides, grows and eventually forms a new individual.

Index

REFERENCES

Curiati, J. A. 2005. Meditation reduces sympathetic activation and improves the quality of life in elderly patients with optimally treated heart failure: A prospective randomized study. *Journal of Alternative and Complementary Medicine* 11(3):465-72.

Dunn, A., M. H. Trivedi, J. B. Kampert, C. G. Clark, and H. O. Chambliss. 2005. Exercise treatment for depression: Efficacy and dose response. *American Journal of Preventive Medicine* 28(1):1-8.

Gangwisch, J. E., S. B. Heymsfield, and B. Boden-Albala. 2006. Short sleep duration as a risk factor for hypertension: Analyses of the first National Health and Nutrition Examination Survey. *Hypertension* 47(5):833-839.

Kamei T., H. Kumano, and S. Masumura. 1997. Changes in immunoregulatory cells associated with psychological stress and humor. *Perceptual and Motor Skills* 84:1296-1298.

Lee, A., and M. L. Done. 1999. The use of nonpharmacologic techniques to prevent post-operation nausea and vomiting: A meta-analysis. *Anesthesia & Analgesia*, online journal of the International Anesthesia Research Society. www.anesthesia-analgesia.org/cgi/content/abstract/88/6/1362.

Myeong, S. L., K. H. Yang, H. J. Huh, H. W. R. Kim, H. S. Lee, and H. T. Chung. 2001. Qi therapy as an intervention to reduce chronic pain and to enhance mood in elderly subjects: A pilot study. *American Journal of Chinese Medicine* 29(2):237-245.

Peng, H. 1990. A primary study on the Han bamboo strip gymnastic book from the Han bamboo strips of Zhangjiashan. *Wenwu* vol. 10: Beijing, China.

Rossman, M. L. 2000. *Guided Imagery for Self-Healing*. Novato, CA: New World Library.

Sancier, K. M. 1996a. Anti-aging benefits of qigong. *Journal of the International Society of Life Information Sciences* 14(1):12-21.

———. 1996b. Medical applications of qigong. *Alternative Therapies in Health and Medicine* 2(1):40-46.

Sancier, K. M., and D. Holman. 2004. Commentary: Multifaceted health benefits of medical qigong. *Journal of Alternative and Complementary Medicine* 10(1):163-166.

Tsang, W. H., C. K. Mok, Y. T. Au Yeung, and S. Y. C. Chan. 2003. The effect of qigong on general and psychosocial health of the elderly with chronic physical illnesses: A randomized clinical trial. *International Journal of Geriatric Psychiatry* 18(5):441-449.

Suzanne B. Friedman, L.Ac., DMQ (China), is an acupuncturist, herbalist, and doctor of medical qigong therapy. Friedman is the first non-Chinese person to be inducted into her teacher's lineage as a Daoist qigong master. She is chair of the Medical Qigong Science Department at the Acupuncture and Integrative Medicine College in Berkeley, CA. Friedman is clinic director of Breath of the Dao, a Chinese medicine clinic in San Francisco, CA. Her articles on qigong and Daoism have appeared in numerous journals and magazines nationwide.

more books from new**harbinger**publications, inc.

EAT, DRINK & BE MINDFUL

How to End Your Struggle with Mindless Eating & Start Savoring Food with Intention & Joy

US $19.95 / ISBN: 978-1572246157

THE HABIT CHANGE WORKBOOK

How to Break Bad Habits & Form Good Ones

US $19.95 / ISBN: 978-1572242630

THE RELAXATION & STRESS REDUCTION WORKBOOK, SIXTH EDITION

US $21.95 / ISBN: 978-1572245495

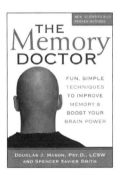

THE MEMORY DOCTOR

Fun, Simple Techniques to Improve Memory & Boost Your Brain Power

US $11.95 / ISBN: 978-1572243705

CALMING YOUR ANXIOUS MIND

How Mindfulness & Compassion Can Free You from Anxiety, Fear & Panic

US $16.95 / ISBN: 978-1572244870

SELF-ESTEEM, THIRD EDITION

US $16.95 / ISBN: 978-1572241985

available from

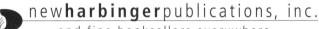

new**harbinger**publications, inc.

and fine booksellers everywhere

To order, call toll free **1-800-748-6273**
or visit our online bookstore at **www.newharbinger.com**

(VISA, MC, AMEX / prices subject to change without notice)